PRACTICAL HOT RODDING

Engineering
STREET RODS

Larry O'Toole

First Published in 1998 by
Graffiti Publications Pty. Ltd.
P.O. Box 232, Castlemaine,
Victoria 3450, Australia.
Phone 61 3 54723653,
Fax 61 3 54723805.
email graffiti@netcon.net.au

© Larry O'Toole, 1998

All rights reserved. With the exception of quoting brief passages for the purposes of review, no part of this publication may be reproduced without prior written permission of the publisher.

The information in this book is true and complete to the best of our knowledge. All recommendations are made without any guarantee on the part of the author or publisher, who also disclaim any liability incurred in connection with the use of this data or specific details.

We recognise that some words, model names and designations, for example, mentioned herein are the property of the trademark holder. We use them for identification purposes only. This is not an official publication.

Graffiti Publications books are also available at discounts in bulk quantity for industrial or sales promotional use. For details write to Graffiti Publications Pty. Ltd., P.O. Box 232, Castlemaine 3450, Victoria, Australia.

Front cover: Chris Henry's T roadster pickup is a perfect example of what can be achieved by the home based hot rodder.

ISBN 0 949398 49 7

Printed and bound in Australia.

CONTENTS

	PREFACE	4
	INTRODUCTION	5
Chapter One	LOOK AND LEARN	9
Chapter Two	CHASSIS AND CROSSMEMBERS	13
Chapter Three	FRONT SUSPENSION AND STEERING	29
Chapter Four	REAR SUSPENSION	51
Chapter Five	MOUNTING THE DRIVELINE	65
Chapter Six	BRAKES	71
Chapter Seven	EXHAUSTS	89
Chapter Eight	COOLING SYSTEMS	97
Chapter Nine	FUEL SYSTEMS	105
Chapter Ten	WHEELS AND TIRES	113
Chapter Eleven	SAFETY ITEMS	119
Chapter Twelve	ADVANCED ENGINEERING FOR STREET RODS	133

PREFACE

Please take a moment to refer to the top line on the cover of this publication. Engineering Street Rods is meant to be used as a guide for the practical hot rodder. The information passed on to you in this publication is a guide only and is based on the author's hands-on experience gained over 30 years of active hot rodding. The author is not a qualified engineer and makes no claim to such qualifications.

In the world of hot rodding there is no substitute for experience. Much of the information included in this publication is from the author's own resources or from the resources of other experienced hot rodders. The information is passed on to the reader in the hope that he or she will be able to apply it to their own street rod project. However all recommendations are made without any guarantee on the part of the author or publisher, who also disclaim any liability incurred in connection with the use of this data or specific details.

This book is dedicated to those street rodders out there who are about to start work on their first project car. The trepidation one feels when undertaking what appears to be a daunting task is soon replaced by a feeling of satisfaction once you finish your first street rod. When I first started out in this hobby there was little in-depth information about in book form so we had to learn the hard way by trial and error. It is my sincere hope that this book will help new and old rodders alike to short circuit that process and eliminate as much of the error as possible.

I would also like to dedicate this book to those established rodders who have helped me in even the simplest way and whose combined experience is incorporated into this collection of information.

Larry O'Toole.

Measurements in this book are in imperial sizes in some instances and metric in others. A conversion table appears on page 137.

INTRODUCTION

Building your own Street Rod is a most rewarding experience. Nothing quite matches the sense of achievement gained from having "done it yourself". Not everyone has the time or ability to do all or even a major portion of the work on their own street rod but it is still important to have at least a basic understanding of the engineering principles involved in building a safe and reliable street rod. That is the ultimate aim of this book, to give the average street rodder a more complete understanding of what it takes to build his or her own street rod, secure in the knowledge that the components used in the car and the way they are fitted is in line with safe engineering practise. Street rodding is a great family hobby, let's do our bit to make it as safe as possible.

This book isn't designed to turn you into a fully qualified engineer but it should help you understand at least the basics of good street rod engineering. Despite the fact that every street rod is different and they all use an assortment of components from various sources there are still some basic design

With the right guidance, plenty of confidence and some devotion to the project, you can take a pile of parts like this and turn it into your own street rod. There are few things in life more satisfying than taking your first drive in a street rod that you have built yourself in your own home workshop.

Above: *There's a real sense of accomplishment in bolting together all the parts that will eventually make up your own street rod. Here I'm bolting back together the rear end in my own Model A Tudor after stepping the rear of the chassis for more suspension clearance.*

Below: *Trial fitting the front sheetmetal on this Anglia means much of the chassis and suspension work is completed. Don't start preparing anything for paint until you're sure it all fits together as planned. This '48 Anglia is Buick V6 powered.*

elements that can be applied to most home-built projects. Traditional styled street rods almost always use some form of the basic original early Ford transverse sprung front suspension system. While it may take several forms, the basic understanding of how it works and what should and should not be done to this type of suspension has wide application. When it comes to independent suspension systems things get a lot more complicated, but if you have at least a basic understanding of how they work you will avoid making unsafe modifications that could jeopardise your safety.

Naturally the same approach to rear suspension systems also applies. It is one thing to fit another rear end into your street rod but there is more to it than simply

taking a rear end from a modern vehicle and bolting it into your early chassis. Things like pinion angle, shock absorber mounting, brake line routing and alignment to the chassis are all important items that need to be taken into consideration. Just a basic knowledge of such items can go a long way to ensuring your street rod is well engineered, reliable in use and essentially safe to use on the road, both for yourself, your family and other road users.

Larry O'Toole.

Right: The rolling chassis for the author's own Model A closed cab pickup goes together for the last time with everything painted and detailed. When you get to this stage it's a good idea to set aside an area of your garage and lay down an old piece of carpet to work on. This will save damage to the detailed components of your undercarriage and it is much more comfortable to work on.

Below: A wide range of ready made suspension and driveline components are available from an ever growing number of specialist manufacturers and suppliers. It has never been easier to build your own street rod than it is now. However all these components still need to be fitted into your street rod in a safe, well engineered manner. If you don't have good welding skills you should either take the time to learn or hand over the major safety related aspects of your project to a skilled tradesman.

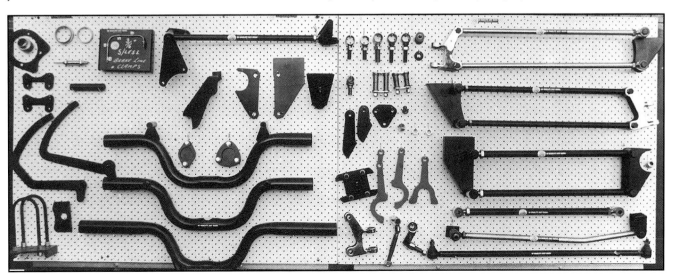

CHAPTER 1
LOOK AND LEARN

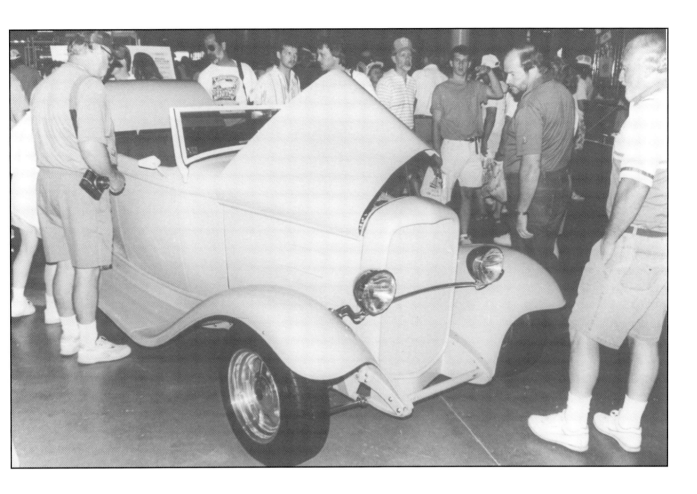

Hot rod shows and rod runs are great arenas of learning. Use them for your own benefit. Before you set about doing something on your own street rod a certain way, take some time to check how others have gone about the same process.

Nothing quite matches the sense of achievement gained from having built your own street rod. Often the best place to learn is to frequent the rod runs and shows where you can inspect at close range other street rods like this neat Deuce hiboy roadster. The right combination of wheels and tires and simple, traditional styling is still hard to beat. Couldn't you see yourself driving your own street rod that looks this good?

Nothing helps you learn faster than hands-on experience. Unfortunately in the process of such an education you will often make mistakes. That isn't something to be afraid of, it is the only way to learn. However you can minimise the number and frequency of mistakes you make by being observant.

Before you set about doing something on your own street rod a certain way, take some time to check out how others have gone about the same process. If you can't find another street rod that incorporates something you have in mind be careful, there could be a very good reason. Read as much as you can on the subject before you fire up your welder and start building your own street rod. Visit and talk to other rodders when you get the chance, ask questions and keep looking all the time. Hot rod shows and rod runs are great arenas of learning. Use them for your own benefit.

Remember that others can do it wrong too. Before adopting someone else's method of doing something, make sure it is soundly based on good engineering principles before you do the same. Even expensive, so-called pro-built street rods can exhibit dubious engineering methods that should not be duplicated, especially for regular street use. Many of these rods are really show cars parading as street rods. Put to the test in real life driving situations their shortcomings would soon become obvious. However that isn't always the case. Look carefully at all pro-built rods and you will soon come to discern the well-built (from an engineering point of view) from the flashy show car that can't be practically driven to any great degree.

In some places the registration authorities require

Before you set about doing something on your own street rod a certain way, take some time to check out how others have gone about the same process. If you can't find another street rod that incorporates something you have in mind be careful, there could be a very good reason. Visit and talk with other rodders to find out if there's a particular reason for doing something the way they did, it may save you time and money in the long run.

that a qualified engineer certify your vehicle before it can be licensed. If this is the case make sure you consult such an engineer before you complete the rolling chassis. That way there is still time to make alterations, if required by the engineer, before you add the paint and detail.

If you are a first time street rod builder I would also suggest you join a club in your local area. While you may feel you won't fit in until you have a street rod on the road you will find other club members can help you immensely along the way and maybe save you from some expensive mistakes as well. Don't hesitate to go to rod runs and shows even if your rod isn't done yet

Hot rod shows and runs are great arenas of learning, use them for your own benefit and try to talk to the owners of cars similar to your own project.

Old tin can be turned into a smart street rod but it needs to be done practically and with safety in mind. The purpose of this book is to help you through that process and hopefully enable you to realise your street rod dream.

either. Exposure to this type of activity will only serve to build your enthusiasm and keep you interested in your project.

Finally I would also suggest you read everything you can relating to the hobby of building street rods. No single book can tell you everything there is to know about street rodding. There are many titles available on the subject and even if you only learn one new thing from each book you buy you will have justified the purchase price. Elsewhere in this book there is a list of suggested reading material to help broaden your street rod building knowledge.

Above all else enjoy building your street rods, this is a hobby not a life sentence, so don't let it become a burden to you or your family. To ensure that always remains the case can I suggest that you consider the most practical ways to build your rod and always consider safety first. Don't underestimate the influence you can have on others either, especially the young. A helpful piece of advice or a ride in your street rod could be just the thing to convert a young auto enthusiast into another fanatical street rodder.

Exposure to rod run activity will build your enthusiasm and keep you interested in your project. Above all else enjoy building your street rods.

CHAPTER 2
CHASSIS AND CROSSMEMBERS

There are some important basic requirements you will need to consider as you set about building your street rod chassis.

THE BASIC CHASSIS

Original or new? For the average home based hot rod builder the obvious choice for the basis of their street rod project will usually be a modified original chassis, especially if you are starting with an essentially stock original car. Look to the reproduction aftermarket and the decision might be different. Here the basis will often take the form of a brand new reproduction chassis (or just a set of new rails) and a fiberglass repro body. Either way there are some important basic requirements you will need to consider as you set about building your street rod chassis.

In their original form most chassis from pre-World War II cars, the origins of most street rods, were fairly flexible arrangements made of folded steel. The chassis of such cars was often designed to flex substantially as a way of compensating for relatively inefficient suspension systems and unduly rough terrain. Now it's a different story. Modern suspension systems are generally much more efficient, but to get the best out of them they really need to be attached to a very rigid sub-structure or chassis. For the average street rod that usually means your original or reproduction chassis will need to be boxed to gain that additional strength and rigidity.

BOXING THE CHASSIS

There are two basic methods of boxing a chassis. Usually the original chassis will be a 'C' shaped section with the open side facing inwards on the chassis. Boxing in this open side adds greatly to the strength of the chassis. To do this you will need to obtain steel plate of approximately the same thickness as the chassis, then cut and shape it to match the profile of your chassis.

Again there are two simple ways of fitting the boxing plates to the chassis before you weld them in place. The first is to simply clamp the plates to the open side of the C section after having chamfered the edges to allow for good weld penetration. The second method is to make the boxing plates so that they are a snug fit inside the edge of the C section. This method is a little more exacting and does leave the chassis rail the same width as it was originally whereas the first method leaves you with a rail that is wider by the thickness of the plate.

Before you start welding make sure your chassis is rigidly supported and levelled in every direction. Clamp it in place so that it can't flex or move during welding and leave it clamped until all welds have cooled completely. Unless you are a competent welder you should have this sort of work performed by a qualified, or at least very experienced person. It

In its original state this swap meet Model A chassis is a flimsy, basic item that lacks both torsional and longitudinal strength, but it could still become the basis of your next street rod project once properly refurbished and strengthened.

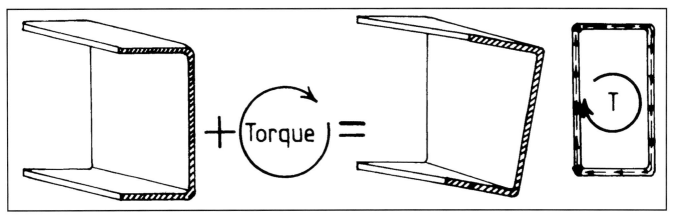

Boxing provides a much stiffer frame section because it greatly reduces the ability of the rail to twist. This open section has a very low resistance to torsion compared with a box section or boxed channel section.

goes without saying that your chassis and the boxing plates must be very clean and free from contamination with paint or grease before welding commences.

If boxing your chassis requires that the original crossmembers be removed, as is often the case, be sure to first tack weld, in fairly substantial fashion, some braces across the chassis so that the original dimensions are maintained. It is common to be able to leave the front and rear crossmembers in place and only remove the original center crossmember when boxing a chassis. The center crossmember usually will require modifications to accommodate late model running gear anyway.

Before you start boxing the chassis you would be advised to weld short sections of steel to the back of all mounting holes for later drilling and threading, since you will no longer have ready access to the inside of the chassis to fit nuts after the boxing is in place. An alternative is to weld nuts to the inside of the holes, but if you do you will need to run a tap through them after welding as the heat will distort the threads.

Depending on the year and make of your chassis it will most likely be a simple ladder frame design with straight crossmembers or it will be shaped to follow the essential design of the lower body and have an X member in the center. If it is of the simple ladder design you will need to box the rails along their entire length and the chassis should be strengthened with the addition of a center K member or X member rather than just a straight cross

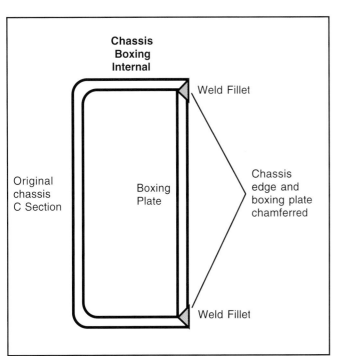

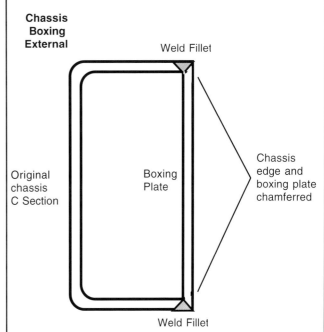

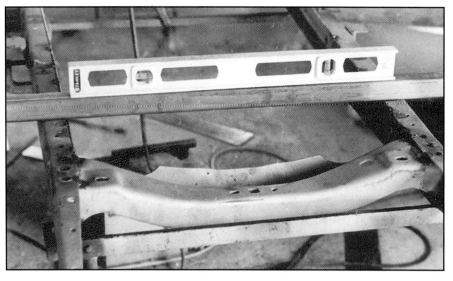

Before tack welding boxing plates into your chassis make sure it is as square and level as possible. Then clamp it securely in position so that it can't move during the welding phase.

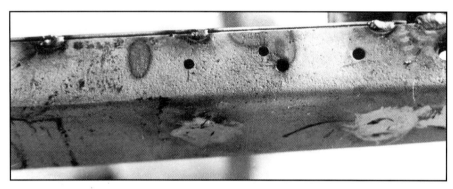

Look closely and you can see that the edge of the chassis rail and the boxing plate has been ground on a taper for good weld penetration. Tack welds are added every three or four inches.

Boxing plates should be securely clamped in place while they are tack welded. The length of angle iron welded across the chassis is temporary to hold the rails in their correct relationship to each other until such time as a center K member or X member can be fitted.

member. The more perimeter shaped chassis that became popular in the mid-thirties can be fully boxed along its entire length (recommended if using high horsepower engines) or it can be boxed forward of and behind the legs of the X member as a minimum.

K MEMBERS

K members are usually fairly simple in design with the straight section welded into the chassis near the middle of the vehicle and two angled legs extending from the center of this crossmember to the rear of the chassis on each side. Alternatively the straight section might be situated towards the rear of the vehicle with the angle legs extending forward. In some

Before tack welding the chassis boxing in place take the opportunity to back all the body mount holes with pre-threaded plates or weld nuts inside the rails. If you use nuts you will need to run a tap through them again as the heat from welding will distort the threads slightly.

Before finish welding anything on your street rod chassis, set it up on solid stands and after levelling it, measure everything for squareness. In this photo the front axle has been blocked into position and squared to the chassis prior to fitting the four bar locating brackets. The chassis boxing and K member is only tack welded in place at this stage.

To align items like this four bar mounting bracket use a length of box tube clamped in place while you tack weld it in place. Note that the chassis boxing plates remain only tack welded in place at this point

Large G clamps hold the engine mounting crossmember temporarily in place in our Model A chassis. Don't finish weld any major chassis components until you're sure everything fits as it should.

Rear suspension locating components should also be tack welded into position before you finish weld any major components. Beware welding on an empty rear end housing like this as it is very easy to warp it through too much heat. One way the home enthusiast can prevent this is by fully assembling the rear end before final welding.

It always pays to make sure the major components are going to fit properly before finish welding mounting brackets and engine mounts. Sitting the body in place allows floor and firewall clearance to be checked and the engine mounts are only clamped in place for the present. Better to find a problem at this stage than after the major welding has been completed.

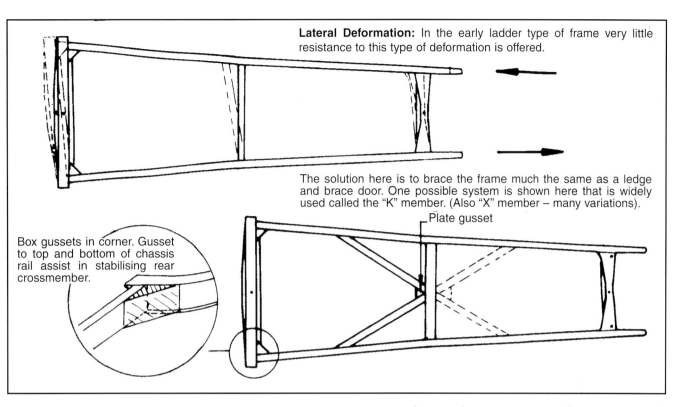

Lateral Deformation: In the early ladder type of frame very little resistance to this type of deformation is offered.

The solution here is to brace the frame much the same as a ledge and brace door. One possible system is shown here that is widely used called the "K" member. (Also "X" member – many variations).

Plate gusset

Box gussets in corner. Gusset to top and bottom of chassis rail assist in stabilising rear crossmember.

This set of diagrams explains the importance of adding K members or X members to early ladder style chassis. The illustrations show a Model A Ford chassis but the same principles apply to any ladder style chassis.

Without sufficient strength added to the center of a chassis it will be deformed too easily by torsional stress at any one corner. A K member or X member is the best way to add the required strength. Particular attention needs to be paid to the design of the center section of such a member.

instances the gearbox mount can be incorporated into the underside of the straight section of the K member or it can be an entirely separate mount. In either case it is a good idea to make the lower section of the transmission mount a bolt-in design for ease of servicing later in the street life of your rod. A properly designed K member will add strength to such a degree that it will prevent your ladder frame from twisting from corner to corner and/or deforming longitudinally.

Study the diagrams that accompany this chapter for a more full understanding of the importance of k members.

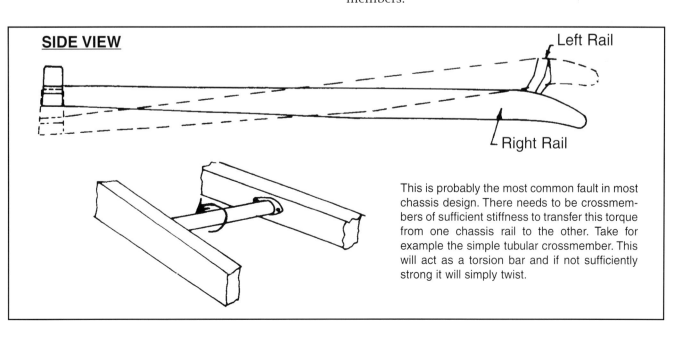

SIDE VIEW

Left Rail

Right Rail

This is probably the most common fault in most chassis design. There needs to be crossmembers of sufficient stiffness to transfer this torque from one chassis rail to the other. Take for example the simple tubular crossmember. This will act as a torsion bar and if not sufficiently strong it will simply twist.

This is where the K or X members are very useful. As the chassis tries to twist the legs of the X act in bending, thus transferring the torque. Remember also that the center of the X is very important to this transfer of torque, you often see a very strong X member rendered useless by lack of connection.

A better system would be to use short pieces of box tubing top and bottom to connect the legs of the X. This also applies to modified X members such as '34 Fords etc.

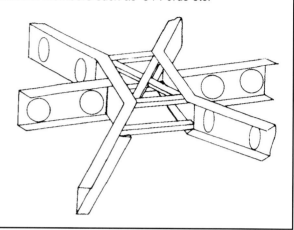

Channel here is unacceptable because it is incapable of transmitting torque

Channel section transmitting torque via bending – OK.

These diagrams show the correct and incorrect ways to make a chassis X member using folded steel channel sections. Strength in the center of the X is paramount.

X MEMBERS

Even original type X members do a reasonable job of maintaining the strength of an early chassis but when combined with partial or full boxing the strength of the chassis is increased dramatically. However particular attention should be paid to the design of the center of your X member as often modifications in this area adversely affect the torsional strength of the chassis. The design of such a center X member should be such that it can still resist torsional twisting but at the same time provide clearance for your updated running gear. Study the diagrams and photos that accompany this chapter to more fully understand this process.

REAR CROSSMEMBERS

Should you decide to retain the standard suspension system in the rear of your early chassis there will be very little you need to do with the rear crossmember. However updating to modern tube shock absorbers will usually require that new mounts be made. A

Compare this reproduction Model A chassis with the flimsy original at the start of this chapter and you can appreciate how much stronger it is. A removable engine and transmission crossmember will also make future maintenance or component swapping easy.

A strong, well built chassis is the basis of any safe and reliable street rod. This reproduction Model A Ford example made from rectangular steel tube has sturdy X member to enhance its torsional rigidity and to resist longitudinal distortion. The rear of the chassis has also been stepped up for greater rear end clearance.

From the rear the reproduction Model A chassis can be studied more closely. In this instance the original kink near the center of the side rails has been eliminated to give a little more space within the chassis. Notice also how the rear crossmember has been kicked rearward to provide extra rear end clearance.

couple of things to watch out for in this area include providing enough clearance between the edge of the crossmember and the body of the shock absorber. You will also need to pay attention to the angle of operation of the shock absorber. If the mounting angle is too great the working efficiency of the shocker can be greatly diminished. As a general rule the shock absorber should be as upright as possible and

Below: *The author built this '32 Ford chassis several years ago. It has been fully boxed and fitted with a custom made X member made from square steel tube. A straight rectangular steel tube rear crossmember replaces the original and a Model A front crossmember is used to lower the ride height at the front by about one inch.*

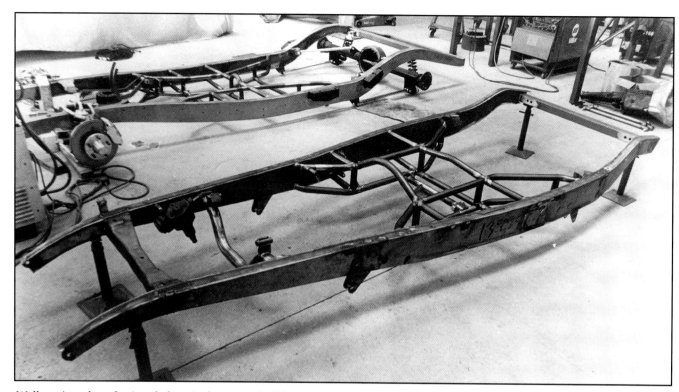

Well equipped professional chassis shops can build a state of the art chassis for you if you don't possess the skills or equipment yourself. Here we have a '32 Ford chassis in the foreground and a '35-'40 Ford chassis in the background. Both have well designed tubular center X members and both have been fully boxed. Sometimes it is a good idea to have your basic chassis built by a professional shop and then you finish it yourself from there. That way you know you're starting with a good foundation.

If you have the space and tools to make one, a complete chassis table is a big asset when constructing a street rod chassis. This one holds a '34 Ford chassis at waist height for easy access while working and provides a basis from which all measurements can be taken. The engine shown here is a fuel injected Holden V8 with five speed manual transmission. Front end is Holden Torana and rear end is Ford.

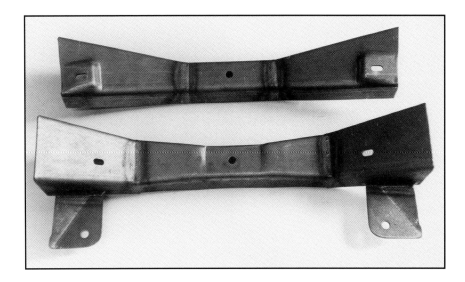

Left: *Early Ford front crossmembers have very little castor built into the front crossmember. Once you build in a little street rod rake the problem becomes even worse. To overcome the situation this crossmember has been cut and laid down at the rear edge so that the spring and axle will have a positive castor of some 6°. Castor in the front end aids straight-line stability in the steering and helps the steering to "self-center" after turning.*

Above: *These reproduction front crossmembers have been manufactured with increased castor built in to suit street rod use. The top one is for a '32 Ford while the lower one is to suit '33-'34 Ford. It is usually better to use a new crossmember like this in your street rod chassis as the originals are often badly damaged from years of abuse.*

certainly never mounted at more than 30° from the vertical.

Replacing the original rear crossmember is sometimes required, particularly if using coil-over shock absorber suspension or if the design of the original makes it hard to incorporate modern shock absorbers or springs. Take care when fitting a new rear crossmember that it doesn't interfere with the floor of the vehicle and if it needs to be stepped upward or backward make sure it is securely gussetted in the corners.

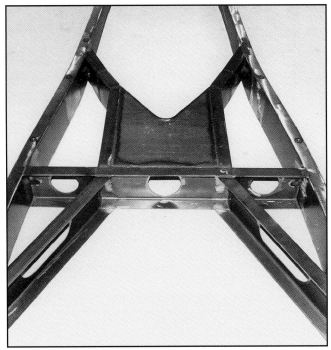

Steel channel can be used for making an X member but it will require extra reinforcement to resist torsional twisting across the center. That's why this example has the large flat plate added to the top of the center section. A bolt-in transmission mount will add further strength to the lower section of the X member.

Even T bucket chassis should have at least a K member incorporated in their design to give them torsional rigidity. Here's a simple design that works well in a confined space and looks attractive at the same time.

Here's a more open style center K member in a Model A Ford chassis. Square steel tube is used to form most of the member and a drop-out transmission mount incorporated into the lower section. Slight rerouting of the near side lower leg also allows a battery cradle to be incorporated. Forward planning provides a mounting point for the brake master cylinder and pedal support bracket.

Another reproduction Model A Ford chassis but this one has more step-up at the rear than usual. This will require the floor to be modified. Note fish plating of welded joints where the step-up has been made. Some authorities require bracing in this manner to provide a larger welded area and consequently less chance of failure. This photo clearly shows the extra strength added by a substantial K member.

A fully boxed chassis also needs a substantial center X member. In this case round tube has been bent to form an attractive X member that will perform well. The open frame design of the X member leaves plenty of room for routing exhausts, brake and fuel lines despite also providing the mounting points for the brake master cylinder and pedal. Note that the transmission mount can be unbolted separately for easy maintenance later in the life of this street rod.

This reproduction '35-'40 Ford chassis has a center X member made from rectangular steel tube to approximate the original design but with a more open center section for transmission clearance. Again the transmission mount has been made removable and the brake master cylinder mount has been incorporated into the design. Be careful when designing center X members like this that you retain strength through the center or you could negate all your fabrication work. If the center of the X isn't substantially braced it will have little torsional resistance, negating its main purpose.

Here's an essentially stock '41-48 Ford chassis that features an array of components that have all been designed to bolt in. Of particular note is the center of the X which replaces the original item without compromising strength, but at the same time it provides more clearance for a later model transmission.

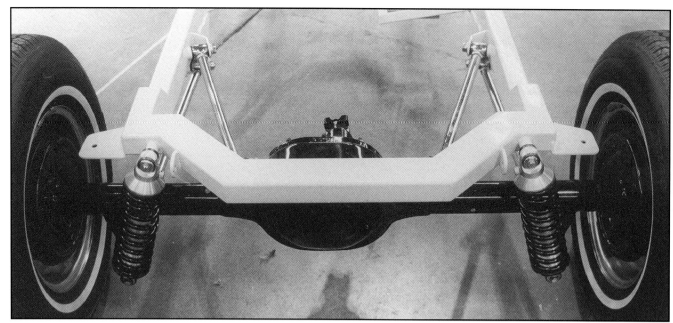

Above: *Stepping the rear of a Model A Ford chassis allows for much needed rear suspension travel but creates a couple of other problems. The rear of the floor needs to be raised if the step-up is any more than two inches and the rails need to be kicked inwards as shown here to bring them inside the lower body rails. Note the rear body mounts that are originally at the outer ends of the chassis have been retained but now they correspond with the lower edge of the chassis step-up.*

Below: *The stock rear crossmember and transverse spring has been retained on this chassis but note how the forward lip of the crossmember has been trimmed to provide shock absorber clearance. However leverage on the upper shock absorber mounts will be severe and they will need better bracing than shown here.*

Below: *One way to avoid having to step the rear of a Model A Ford chassis is to widen the rear crossmember as shown here. This allows a later model rear end to move up into the rear crossmember with less chance of hitting the lower edges. The chassis needs to be shortened slightly to let the rear end center under the crossmember again. This chassis has a simple but strong round tube K member fitted as well.*

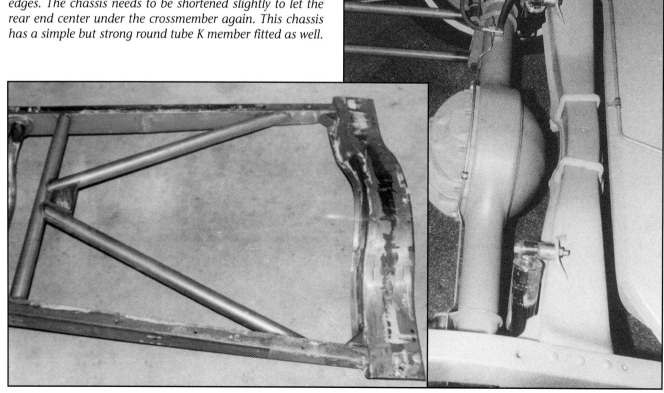

Some smaller cars like the English Ford Anglias and Prefects have a fairly light chassis that isn't easily strengthened in the conventional manner. These chassis are better suited to having a second chassis built within the original so that the new chassis carries the mechanical components and the original rails basically only serve to locate the body. That's what has been done to the chassis shown here which also has a substantial K member added. Notice how the semi-elliptic spring mounts are tied into the new and old chassis adjacent to the kick up.

Here's a side on view of an Anglia chassis fitted with a new internal chassis to give it the necessary strength to support more powerful running gear. This one has been fitted with Holden Torana front and rear suspension. Rebuilding one of these chassis in this manner also provides more side impact protection as the original rails are supported by and connected to the new internal "T bucket Style" chassis.

Final assembly time of your painted and detailed chassis and running gear is a satisfying time. An old piece of carpet on the garage floor keeps everything clean and prevents damage to painted components as the reassembly begins.

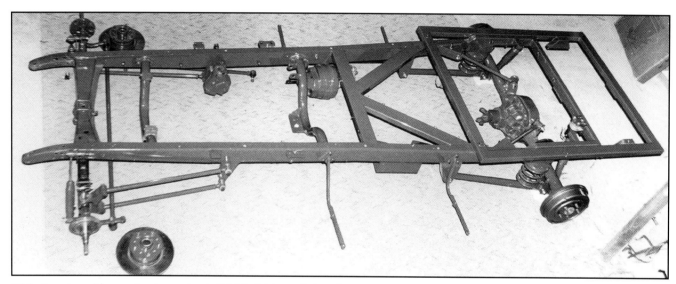

This chassis resides under the author's '30 Model A Ford closed cab pickup which was completed in 1990. The separate frame at the rear mounts the pickup bed and it bolts in place.

The finished, rolling chassis is pushed out into the yard for the first time. Running gear is Buick V6 with Turbo 350 auto transmission. Rear end is eight inch Ford located by an unequal length four link with coil springs from a Holden Torana, an inexpensive suspension system that works well.

CHAPTER 3
FRONT SUSPENSION & STEERING

Whether you use independent suspension or a beam axle front end it is vital that you set it up correctly.

TRADITIONAL EARLY FORD TYPE FRONT ENDS

In its original form the traditional early Ford I beam front axle with locating wishbone and transverse spring is simplicity itself. Yet for all its simplicity it can work quite efficiently. Just making sure all the components are in good working order and adding some modern shock absorbers can make the beam axle front end at least serviceable for a budget conscious rodder. However the use of late model running gear in a street rod that originally used an early Ford wishbone front end usually means that the pivot mechanism near the center of the vehicle is located in the wrong place. The traditional solution to this dilemma has been to split the arms (or radius rods) and attach them to the chassis rails at each side. This provides the required space for the later model running gear but it also introduces some other problems.

Mounting the ends of the radius rods via tie rod ends or urethane bushes (preferable) to the chassis rails immediately puts the whole front suspension into bind because it no longer pivots around that central ball that it used originally. If a traditional style I beam axle is retained the system can work moderately successfully because the design of the I beam axle actually allows the axle to twist along its length, giving the front end some degree of flexibility and allowing the suspension to still work. However add a round tube axle and this system goes into total bind. The suspension system can now essentially only work in one plane, up and down equally on both sides. Don't use a system like this in your street rod as it is inviting dangerous handling in extreme situations. If you can't at least retain an I beam axle look at substituting a four bar system for your split radius rods. It's the only way to gain a front suspension system with tube axle that actually works properly.

FOUR BAR SUSPENSION SYSTEMS

The use of a four bar front suspension system overcomes the problems outlined above when radius rods are split. By using two bars on each side the axle can now move in a vertical plane as the suspension works up and down. The use of compliant

Original early Ford front axle and wishbone is as simple and basic as suspension systems can be, yet it can be made to work quite efficiently. Dropping the center section of the axle is a traditional method of lowering the ride height of street rods equipped with this type of front end but it wasn't always done as neatly as this one. The axle ends should not show signs of severe distortion or stretching. This '32 Ford axle was dropped by GMK Engineering in Australia and it comes with engineer certification and identification number.

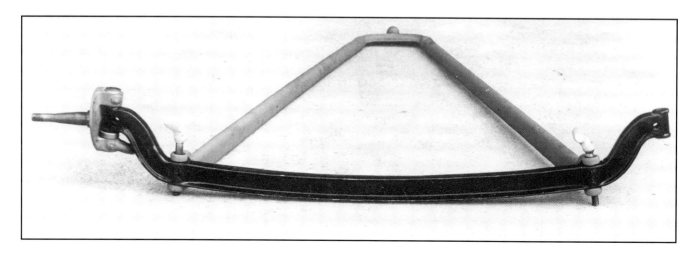

The diagrams shown on this page clearly depict the effect that suspension and steering layout can have on the driveability of your street rod. Where the arcs differ greatly as in the middle diagram, bump steer will be experienced and you will feel it at the steering wheel. Note that tube axles should not be used with split radius rods as the axle must be able to twist for this type of suspension to work satisfactorily.

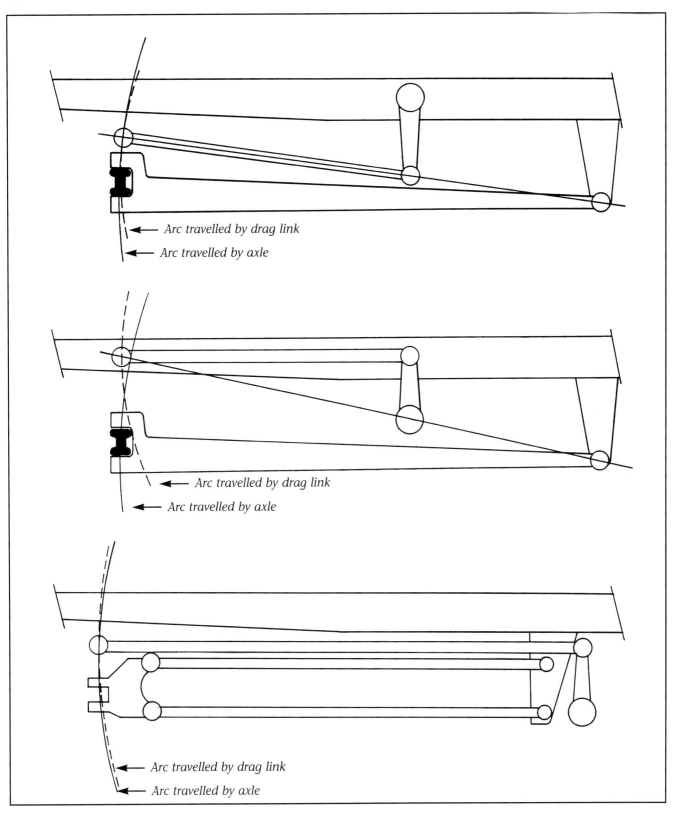

A severely dropped axle like the one shown here will really get your street rod in the weeds but beware, this axle and its associated lower shock absorber mount will come into contact with the road before the wheel rim does in the event of a blowout or sudden tire deflation. Either use an axle with a less severe drop or fit a larger diameter wheel so that the lowest point of the vehicle becomes the wheel rim

The scrub line on this T bucket front end is better in that the wheel rims are the lowest point. However a broken spring or shackle on this aptly named suicide front end would allow the chassis to drop onto the road with disastrous results. Some sort of fail-safe restraining device needs to be fitted.

Here's how simple that restraining device can be. If the spring or a shackle on this front end breaks the chassis can only drop until the simple licence plate bracket comes into contact with the axle. The vehicle could probably even be driven at slow speed to a place of safety or repair.

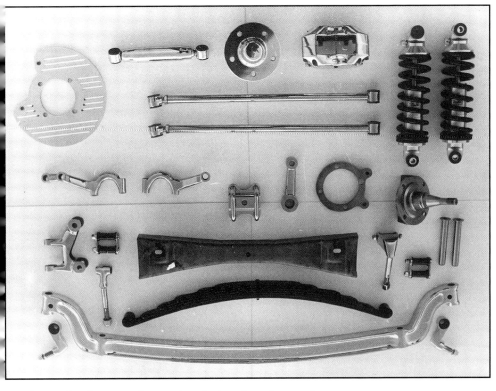

Street rod suspension can be as simple or as exotic as you want. Shown here is a complete range of front suspension components as supplied by Rod City Repro's in Australia. Most parts are available in either stainless steel or mild steel and priced accordingly.

urethane bushes at the ends of the four bars also allows the system to twist slightly, thus giving a much freer suspension. As the four bars move up and down they do actually trace an arc about their rear pivot point and this results in the axle actually being pulled slightly towards the rear of the vehicle. However this isn't noticed in the handling as the amount of movement in a vertical plane that the axle traverses is in fact quite short. The longer the four bars are, the less rearward movement there will be.

FRONT SPRINGS

Now that we have an axle that can move freely we can turn our attention to the front spring and its mounting crossmember. If you are going to get the best out of an early Ford type transverse front suspension system you need to ensure that the spring itself is in perfect working condition. You may need to

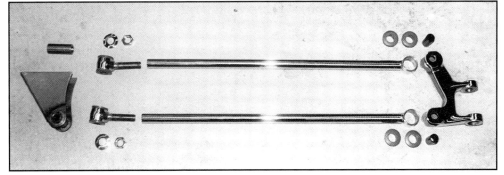
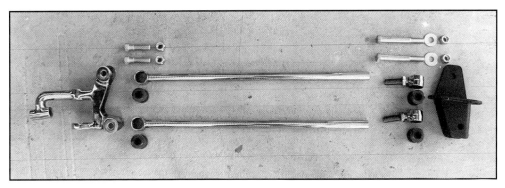

The most popular method of locating a beam axle front suspension system is with a well designed four bar arrangement. The upper photo shows a four bar system to suit '28-'34 Ford front ends with the spring located above the axle while the lower example is for '35-'48 Fords where the spring is located in front of the axle. Both these examples are made from stainless steel but mild steel versions are available for the more budget conscious rodder.

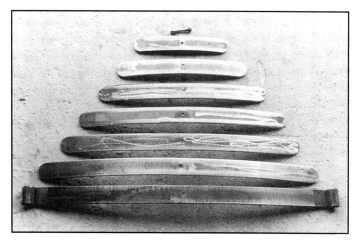

One of the secrets to improving the ride with an early Ford transverse type suspension system is to give the spring itself a "tune up". Completely disassemble the spring so that you can work on each leaf individually. Thoroughly clean each leaf and use a linisher to give each leaf a uniform, rounded shape. Use the linisher to also taper the underside of the ends of the spring leaves so that they will have a tendency to slide over each other rather than dig into the surface of the next leaf down. Even new springs can be quite rough at the ends so don't just assume that because your spring is new it will work efficiently.

take the spring apart, even if it is brand new, thoroughly clean it and before reassembling it slightly chamfer the ends of the leaves on a linishing belt so that they will slide over rather than dig into each other. Thoroughly grease the leaves before reassembly or fit teflon between the leaves. Some manufacturers even supply springs that have teflon buttons incorporated into the end of each leaf to allow them to slide easily over each other. If you grease a spring you will need to wrap the spring to keep the grease inside. A very effective alternative which I have used myself is to cover the spring with electrical type shrink tubing. Once heated and shrunk into place it does a very good job of retaining the grease and it looks nice and tidy to boot.

Don't mess around with old perch pins and shackles when building your transverse front end. New parts are easily obtained and the modern technology used in shackle bushes is far superior to the old fashioned bushes that required greasing.

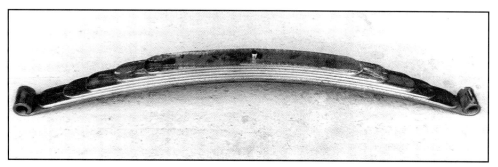

Having smoothed, rounded and tapered the underside of each leaf add grease between each leaf as the spring is reassembled. This one is looking better already.

Your "tuned up" spring will now work quite nicely but you need a way to keep the dirt out and the grease in to maintain its ride quality. They can be wrapped with tape but a better way is to slide on electrical shrink tubing from each end of the spring and heat shrink it so that it encapsulates the spring but for the very center section and the eyes at the outer ends. Here the heat from an oxy-acetylene torch is being used to shrink the tubing but care is needed so that you don't burn the plastic.

FRONT CROSSMEMBERS

Next we need to turn our attention to the front crossmember. In most early Fords the front crossmember is relatively flat where the spring locates. Street rodders tend to run a little extra rake in their chassis so it is important to compensate for this when setting up your front suspension. With the chassis sitting at its final ride attitude you can measure and adjust the center portion of your front crossmember so that the spring mounting surface is at 6-7° positive castor (sloping towards the rear). This castor in the front end will allow the car to track straight ahead better and it also helps the steering to self-center after turning. Some aftermarket manufacturers actually incorporate this amount of castor in their reproduction front crossmembers but it still pays to check it out as you build your chassis.

Above: *Once the shrink tube is fitted and trimmed the spring will give long service life and remain flexible in operation thanks to the grease trapped within and the smooth spring leaves which can now slide easily over each other. The finished spring also looks attractive and will remain that way for a long time.*

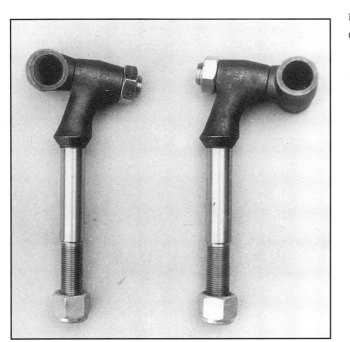

Left: *Adjustable spring perches can overcome the lack of castor in a transverse suspended front end. Castor can be adjusted into the axle while leaving the spring in a more horizontal position. Care still needs to be exercised to ensure that the spring isn't twisted and put into bind when adjusting the axle castor.*

Many important safety factors have been incorporated into this traditional transverse sprung front suspension system. The crossmember has castor built into its raised center (which lowers the ride height) and a plate that bolts to ears on the crossmember eliminates the need for U bolts to retain the spring. Cross steering is used and it has been well set up to keep the drag link and tie rod parallel. An important addition that should be used with all cross steering applications is a panhard bar. This should always be mounted to the chassis on the same side as the steering box and on the axle assembly at the opposite side as shown here. Right hand drive chassis should be the opposite way round, again with the panhard bar mounted to the chassis on the same side as the steering box.

A "dead" perch is simply a spring perch without a shackle that fits to one side of a transverse front spring. This works in the same manner as a panhard bar but uses one half of the spring itself for that purpose. Generally not quite as efficient as a properly engineered panhard bar but better than no lateral restraint at all when a cross steering system is fitted. Ideally a dead perch should be fitted to the passenger side of the front axle, making the example shown here correct for a RHD car.

CROSS STEERING WITH TRANSVERSE SPRING FRONT SUSPENSION SYSTEMS.

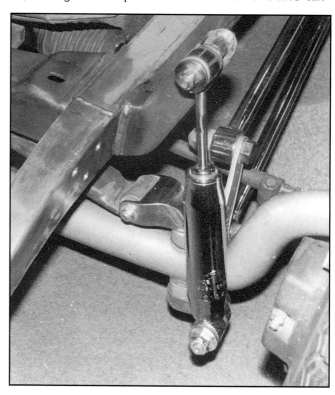

While cross steering is fairly simple to incorporate into a beam axle front end it does have some important considerations too. The drag link should be as close to parallel with the tie rod as is possible at final ride height and it should usually be as far forward as possible without creating binding between drag link and tie rod.

Without doubt the single most important factor to consider when setting up cross steering is to allow for the jacking affect produced by the drag link pushing in one direction on the spindle mounted steering arm. The shackle system that allows the freedom of

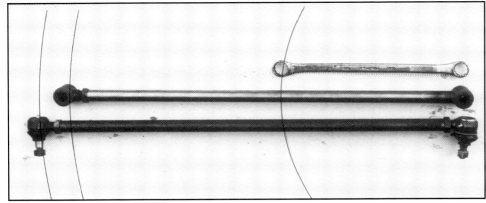

Left: *This practical example shows why a panhard bar should be as close to the same length as the drag link as possible. The Panhard bar in the middle transcribes a similar arc to the drag link when their lengths are similar. The wrench serves to illustrate how the change of arc caused by a much shorter panhard bar would have an adverse affect on steering.*

Right: *Here's a good example of a front mounted panhard bar on a '32 Ford street rod. This is a right hand drive vehicle so note that the panhard bar mounts to the same side of the chassis as the steering box and to the axle on the passenger side of the car. Left hand drive vehicles should be opposite to this. The panhard bar should be as close to level with the axle in the final riding position as is possible and parallel to the drag link.*

movement needed in a beam axle front end will also allow the drag link to push the axle or chassis to one side on the shackles. Conversely the effect of a wheel hitting a bump on the road can be transferred into the steering drag link and cause the steering to turn just when you don't want it to, like in the middle of a high speed cornering manoeuvre.

The answer to this problem is to provide some sort of restraining device that will let the suspension pivot on its shackles but negate any sideways movement. The most common way to achieve this result is to fit a panhard rod between the axle and the chassis. This should always mount to the chassis rail on the same side as the steering box and to the axle on the opposite side. It should also be as close to the same length as the drag link as is possible and again be parallel with it.

Another method of achieving a similar result is to fit a dead perch to the passenger side of the vehicle's front suspension. In effect this allows one side of the transverse spring to act as a panhard rod while the conventional shackle on the opposite side still allows the suspension to move up and down in a flexible manner. This system is something of a compromise as the half spring panhard bar is really too short and therefore doesn't move through the same arc as the drag link during suspension movement. The fact that suspension movement in this type of front end is limited generally means that it may not be sufficient to create major problems but I still favor the full panhard system where it is possible to fit same.

Other central locating devices can be manufactured for these types of front ends but they are generally more complex than we wish to cover in this book. For the great majority of transverse suspended, beam axle equipped street rods either of the two methods of incorporating and restraining cross steering explained here will be most suitable.

We showed you this chassis in an earlier chapter where we pointed out the cleverly designed bolt-in center X member brace. This allows a later transmission to be more easily fitted but to do so the stock early Ford wishbone needs to be split and the ends mounted to the forward legs of the X member as shown. To preserve the integrity of the original concept the ends of the split wishbone still need to be kept as close as possible to the center of the vehicle and the I beam axle should be retained. Also note the original style pedal mount that has been made to accept a modern dual circuit brake master cylinder.

Here's a complete custom front suspension system for an early Chevy chassis that incorporates a dropped tube axle mounted on semi-elliptic springs and incorporating a cross steering system. All components have been designed so that they bolt in, making installation easy for the home based enthusiast. Semi-elliptic springs don't require a panhard bar.

Side steering with a four bar suspension system should be set up like this one. Note how the drag link is as close to the same length as the four bars as is possible. The drag link should also be parallel with the four bars so that all three bars trace the same arc when the suspension moves up and down. This way bump steer will be minimised or eliminated altogether. The steering box used in this right hand drive vehicle is from a Ford Falcon and has been mounted on its side. Early Mustang steering boxes can be used in similar fashion for left hand drive side steer street rods.

SIDE STEERING WITH TRANSVERSE SPRING FRONT SUSPENSION SYSTEMS

Dropped axle transverse spring front ends with four bar locating systems aren't necessarily restricted to early Ford applications. This one has been fitted to an Austin A40 pickup. Look carefully and you will see that it also has front steering that retains its ackerman geometry and includes a front mounted panhard bar.

Sometimes known as push-pull steering, side steering is one of the simplest methods of setting up a steering system in a street rod. Usually fitted in conjunction with a four bar suspension system, the side steering arrangement does have some basic requirements. In particular the steering box should be mounted so that the drag link remains essentially parallel to the four bar arms. It should also be as near to the same length as the four bar arms as possible. This will leave the arms and the drag link working through the same plane which will minimise bump steer.

One thing to keep in mind if you are fitting a dropped axle into an existing front suspension system that formerly used a stock axle is to remount the steering box so that the drag link maintains its relationship to the four bar arms. If you don't do so the drag link will now be running uphill from its former position and bump steer will almost certainly result. The same basic theory also applies to side steering in relation to an original Ford wishbone equipped front end, i.e. keep the drag link moving in the same arc as the wishbone wherever possible.

INDEPENDENT FRONT SUSPENSION.

No matter how well you set up your early Ford transverse front suspension system it will rarely be as good as a properly engineered independent front end. However even independent suspension systems can handle badly if some basic engineering requirements aren't taken into account. Sometimes the most elaborate independent front end is also a nightmare when it comes to ride and handling.

The safest way to go when designing an independent front end is to use a tried and tested system from a late model car or copy its design elements when building your own. Of course some of

Setting up an independent front suspension system correctly is imperative to having it work properly. This one is the popular Mustang II type that has been fitted to an early Chevrolet chassis using a bolt-in mounting kit from Chassis Engineering that includes crossmember, steering mount and stabiliser bar mount. If you don't possess adequate welding skills this is a good way to incorporate such components without compromising the safety of the vehicle.

Another Mustang II front suspension system but this time using a weld-in crossmember from Chassis Engineering to suit '33-'34 Dodge/Plymouth applications. If you use a weld-in crossmember be sure to allow for the final ride attitude of your chassis (rake), don't just assume that if you mount the front end parallel with the bottom of the chassis it will be in the correct position. If necessary take measurements from the original front end donor vehicle (in this instance a Mustang II) and duplicate those measurements in your street rod chassis.

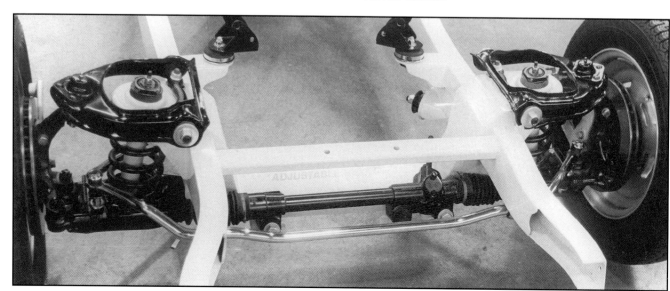

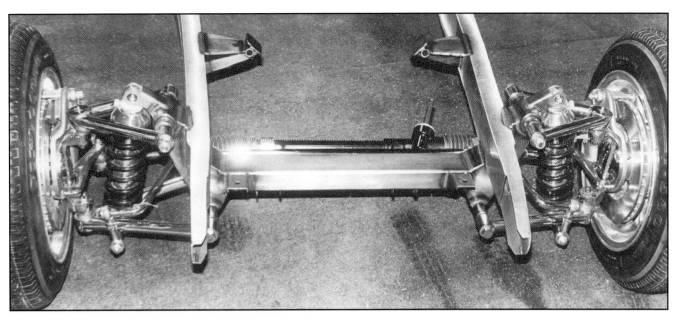

the earlier body styles popular in street rodding have limited space under the front fender so there may be a need for some compromise. This is where extreme caution is required.

Many of the expensive aftermarket independent front ends have been designed to fit under the front fenders of these early cars and provide correct suspension geometry at the same time. Beware however, because some do not have correct geometry.

The main aspect of poorly designed independent front ends is poor leverage angles on the spring or coil-over shocker, that does all the work. To fit it under the fender the upper mounting point is sometimes moved too far inboard and too low so that when the suspension has to work through its complete arc of operation it actually loses some of its leverage advantage as it compresses. In effect this

Custom made independent suspensions are readily available from the hot rod shops but not all are well designed. Beware such suspension systems that don't have the springs mounted as well as this one has. Note how the coil-over shock absorber unit is near vertical to allow the spring to work as designed. If the coil-over is mounted at too great an angle ride quality is compromised and stiffer springs are needed, negating the reason for going to independent suspension in the first place.

This '35-'40 Ford chassis features the exotic independent front suspension from a Corvette. The original dimensions of the Corvette application need to be incorporated into the early Ford chassis to ensure that the suspension works properly. This application will also require modifications to the inner fenders for clearance.

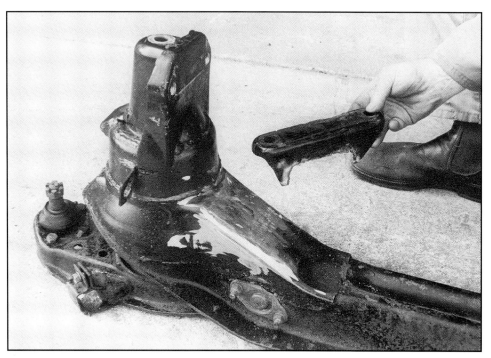

Left: *Usually the stock mounts will need to be modified or replaced when fitting a front suspension crossmember into your street rod. Shown here is the stock mount off a Mitsubishi L300 front end. Where possible try to duplicate the method used by the factory or improve on it when making your own mounts.*

Below: *Here's the same type of front end with angle brackets welded into the crossmember so that it captures the chassis from the side and bottom rails. The brackets have been let into the crossmember to lower the ride height of the vehicle. Rack and pinion steering has been added.*

Left: *Several aftermarket manufacturers now offer dropped stub axles for independent front ends. These are often ideal for street rods since they allow the vehicle to ride low to the ground without sacrificing ride quality. Shown here is a stock Australian Holden stub axle on the left with a two inch dropped version as made by the Castlemaine Rod Shop.*

Chassis Engineering make this torsion bar front suspension system for early Ford chassis. Advantages are clean appearance and adjustable ride height but it must be carefully set up to work well. Modifications to the inner fenders will be required with this type of suspension.

means the suspension gets softer just when it should be offering more resistance. The only way to overcome this, short of redesigning the front end is to fit firmer springs. The result is a suspension system that should be really nice but doesn't ride that well at all. Might as well have retained the transverse spring suspension and not paid out good money for an independent front end that doesn't work properly!

The way to prevent this situation from arising is to ensure that the upper spring mount is outboard of the inner pivot points of the top arm and as vertical as possible. Now the spring can do its work properly and you should get that nice ride.

If you are using the entire independent front suspension system from a donor car but you're not sure how it should be mounted you will always be close to the mark if you duplicate the mounting methods and measurements of the original car. That means you will need to allow for such things as the rake you want to build into your street rod so be sure to take that into account.

The ultimate in independent front suspension is the fully independent unit with stainless steel tubular A arms, alloy coil-over shock absorber and custom made hub with four spot disc brakes. This type of front suspension system is entirely hand made in jigs so it doesn't come cheap but if appearance is important it's hard to beat. Note the adjustability built into the top arms for setting front end alignment.

Two universals are used here to align the intermediate shaft with the steering box and the bottom of the steering column. Note that the universals are "in phase". That is they are aligned to each other along the shaft.

STEERING BOXES

We have already covered using steering boxes in cross steer or side steer applications with beam axle front suspension so now let's take a little time to look at independent front end steering systems. When using a steering box, drag link and idler arm set up in this instance there is much to consider. In fact unless you are very experienced with this aspect of building street rods you should try to keep your steering system as simple as possible. The obvious thing to do is retain the steering system that came with the donor front end if that is possible and be sure to mount it in the same relative position as it was in the donor vehicle. Sometimes that will involve minor chassis alterations to accommodate the positioning of the steering box mount and idler arm mount but if such modifications would mean major changes to the chassis, perhaps it would be better to consider a different steering system.

Another example of a well engineered steering connection. Universals are "in-phase" top and bottom and there is plenty of clearance between steering shafts and engine components. The rear exhaust outlet will be fairly tight but it is better to get the steering in place first and fit the exhaust around it.

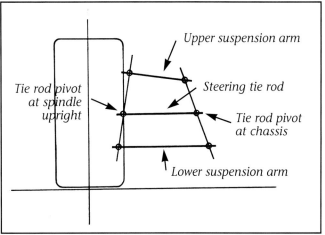

Whenever there is more than one intermediate shaft in your steering linkage a self aligning bearing of some sort will be required. This one has been built into the engine mount where it will be largely hidden once the engine is in place. The Holden front end has been braced and mounted to the chassis in a manner that will withstand torsional forces well. Rack and pinion mounts are also well designed but still look part of the overall package.

Above: *Finding the right mounting position and length of tie rod for the steering on an independent front end can be tricky. This diagram will help you to understand the basics of getting it correct, however there are many other factors that can influence the outcome, such as position and shape of the steering arm. Before you attempt to mount a steering system in your street rod I strongly urge you to seek out more in-depth information on the subject or consult a qualified and experienced engineer. Books such as "How to Make Your Car Handle by Fred Puhn and "Chassis Engineering" by Herb Adams are two good reference books that cover this subject.*

RACK AND PINION

Usually the alternative steering system on an independent front suspension will be a rack and pinion. Again there are some important considerations to be dealt with. It might seem obvious to simply mount the rack and pinion on the rear or front of the crossmember but first you need to do some thorough checking.

Very rarely will a rack and pinion steering system from a third party donor vehicle be suitable for attaching straight to your crossmember without needing to be narrowed to suit the suspension pivot points. This is where the inexperienced rodder can get into serious trouble but using some common sense and a lot of measuring can set you along the right track.

If your rack and pinion steering arm pivots don't line up with a line drawn from top inner suspension arm pivot to lower inner suspension arm pivot you will need to shorten it accordingly so that they do line up at this point. This is a little simplistic as there are other considerations that can come into play but for the most part following this plan of action will prove successful. Most often the rack will mount close to a horizontal position that corresponds to the lower suspension arm and the steering arm pivots will be close to aligning with the lower inner suspension arm pivot.

This is an idea that comes in handy when the steering arm on your early Ford stub axle is missing its second tie rod hole. A short section of a drag link has been used that incorporates a second hole in its length. This allows the drag link to mount at this point instead of directly to the steering arm as would normally be the case.

Here's a close-up of a drag link/tie rod fitting like that shown above. The fitting has a normal taper to accept the tie rod end in the same manner as a steering arm.

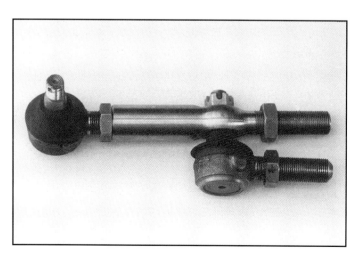

Before undertaking a steering conversion like this you should take the time to study books that deal with the subject in depth and talk to other rodders with experience in this type of work. Shortening a rack and pinion isn't a job for the backyard operator either. Spend a little extra and have it done by a reputable machine shop.

SHAFTS AND UNIVERSALS

Linking the steering box or rack and pinion to the steering column will almost always involve some combination of shafts and universals in order to align the two components with each other and to clear mechanical components at the same time. While this may seem to be a straight forward operation there are some principles to be incorporated if your steering system is to be as safe as possible and operationally efficient.

Steering shafts should never be made from mild steel. Always use a bright steel or if you don't have access to reliable supplies of suitable material at least try to use factory steering shafts that can be modified to suit. Unless there is no way around it you should avoid cutting and welding steering shafts at all times. If it is unavoidable make sure you incorporate some kind of fail-safe mechanism while you're at it.

Usually sections of the steering shaft will be linked together by universals. Only use universals that have been manufactured specially for this purpose in your steering system. Your own safety and that of your passengers is at stake. These universals are usually retained by some type of spline and clamping system or in some instances there will be a flat section on the ends of the shafts with a retaining bolt or cotter pin that engages across this flat to maintain positive contact.

It is important that steering universals aren't fitted at severe angles or they will wear quickly and could fail. Keep them at less than 27° if possible. Sometimes it will be necessary to incorporate an extra shaft to reduce the angle of the universals and where this is the case and you end up with one

The use of a dropped axle with an early Ford wishbone will sometimes mean the steering drag link and tie rod need to pass above the wishbone rather than under it. If this is the case you need to check suspension travel carefully as the steering links could foul against the bottom of the chassis at full suspension travel. Wishbone or chassis rail may need to be modified to provide clearance or the whole system may need to be altered to suit.

Most importantly they need to be in phase with each other. That is the universals should be aligned along the length of the shaft so that they don't "fight' each other in operation.

Street rods often have high horsepower engines and indirectly this can lead to a dangerous situation if you don't provide adequate clearance between the steering shafts and the engine block or headers. You should aim for at least 1/4 inch clearance from any engine components, more is better. Under hard acceleration a high horsepower engine can "torque over"

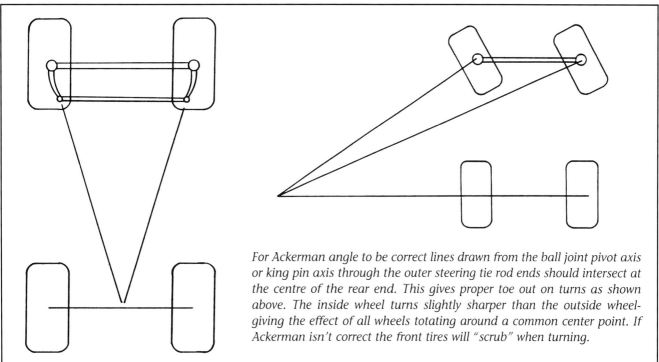

For Ackerman angle to be correct lines drawn from the ball joint pivot axis or king pin axis through the outer steering tie rod ends should intersect at the centre of the rear end. This gives proper toe out on turns as shown above. The inside wheel turns slightly sharper than the outside wheel- giving the effect of all wheels totating around a common center point. If Ackerman isn't correct the front tires will "scrub" when turning.

section of shaft that isn't supported by either the steering box or the column, you will need to incorporate some form of extra support. This usually takes the form of some type of self-aligning bearing that holds the shaft in alignment with the upper and lower shafts. Check the photos for a clearer explanation of this aspect of setting up steering systems.

As with driveline universals the alignment of the steering universals themselves is also important.

enough to jam the steering shafts against such components with obvious disastrous results.

Another steering component that is often misused is the fabric insulator that is found at the steering box where there is sometimes also a universal but not always. This insulator is there to absorb vibration not to act as a universal and it usually incorporates some sort of fail-safe drive method. This is incorporated so that should the fabric insulator fail you will still have

Here's a fiberglass bodied street rod that has been set up well to accept a collapsible steering column and incorporate such modern conveniences as air conditioning. A square steel tube brace runs right across the lower dash area which provides a secure anchor point for the steering column mount which includes breakaway slides.

steering. Don't eliminate this positive drive component from one of these insulators if you need to modify one in your steering system. Also avoid using a fabric insulator as a substitute universal. They are not designed to work with anything more than slight misalignment of steering components and should not be stressed at all. Again, study how they are fitted and operate in factory built mass production vehicles and copy their methods.

MOUNTING STEERING COLUMNS

Few street rodders would not be conscious of the need to incorporate as many safety items into their street rod as they are able. Things like seat belts, neutral isolator switches, third brake lights and collapsible steering columns are all easily incorporated into street rods with a little advance thought. However when it comes to collapsible steering columns I see many street rods where the lower part of the collapsible system is incorporated but the upper is not. Let me explain.

Most collapsible columns have a sliding shaft arrangement in the lower section that is held in place with small spot welds or some similar retaining method that will shear in the event of an accident and allow the lower section of the shaft to slide up inside the upper section, thus absorbing much of the impact and preventing it from being transferred all the way up the column in spear like fashion. However there is a secondary function in the modern collapsible column that is often overlooked by street rodders in the mounting system of the upper section of the column.

Usually the upper steering mount bracket on the column is bolted solidly in place on the column tube, but the bracket itself is in turn bolted to the lower section of the dash in a special way. Look at one in a modern car and you will see that the bolts pass through small alloy blocks that slide into the steering column bracket in such a way that sudden impact on the steering wheel, i.e. the driver hitting it in an accident, will allow the upper section of the column to collapse downwards. In this instance the alloy slides stay where they are and the whole upper column depresses towards the firewall in a controlled manner. Too often we see the upper section of a collapsible column solidly mounted to the dash,

Right:
When mounting a collapsible steering column keep it that way by incorporating the breakaway slides in the dash mount. Here a separate bracket has been made to accept the slides on a stock Toyota column mount and mate them to the original Model A Ford column mount. A simple plastic or light aluminum cover can be made to hide the mounts at a later date.

negating half of its ability to collapse during a severe crash situation.

It's not hard to re-engineer one of these mounting systems so that the bracket and slides are less obvious. In fact examples of how it can be achieved are shown in this chapter. If nothing else, mounting your steering column so that it is still completely collapsible will give you greater peace of mind.

Left:
Here's the same steering column with the light aluminum cover fitted. This cover was actually made from a store bought lightweight aluminum baking dish. The plastic fan control switches are all that holds it in place and would break away easily in an accident situation.

CHAPTER 4
REAR SUSPENSION

For the average street rod the conventional live axle rear end is about the simplest way to go.

Typical four bar rear suspension system with coil-over shock absorbers is simple and strong. Note that the leverage forces on the shock absorber eyes and their mounts are very high in this type of installation so they need to be well designed and reinforced to withstand such forces. Top mounts have been gussetted to the rear crossmember while the lower mounts are incorporated into the rear of the four bar axle mount and a machined spacer is used to prevent the coil-over from coming into contact with the rear end during operation.

LIVE AXLE

For the average street rod the conventional live axle rear end is about the simplest way to go. Whether suspended by an original transverse spring early Ford type suspension system or the more popular semi-elliptic springs as found on most modern cars without independent suspension, the basics of setting up this type of rear end are essentially similar. Mounting the rear end square in the chassis isn't too hard. Just measure in several different planes so that the wheel mounting flanges of the axles are the same distance from each side of the chassis and at right angles to the centerline of the chassis itself. What is more important is to get the pinion angle set correctly. If you don't the driveshaft will continually wear out universals.

The secret to setting the pinion angle of a live axle rear end is to rotate the housing until the face of the pinion is at the same angle as the vertical face of the rear of the transmission extension housing. This ensures that the universals will be working in the same planes and therefore binding is eliminated.

If you are retaining a torque tube early Ford type rear end you should also retain the original method of locating the rear end in the chassis. That is by wishbones attached to the torque tube itself and by the stock spring mount in the chassis. Adding modern shock absorbers is an easy installation and well worth the effort. Should you be using an open drive rear end with transverse spring then you will need to be more careful with the method used to locate the rear end in the chassis. A four bar system is the simplest, modern way to locate such a rear end set up but you might also consider using the early Ford type radius rods as used on '35-'36 model Fords that bolt to a rear end housing mounted bracket. If you use this method be sure to fit a substantial bushing at the forward end, similar to that used in

Left: When setting up a rear end suspension system find the required ride height for the rear end housing and then tack weld it in place as shown here after making sure it is square in the chassis. Now you can assemble all the components of your rear suspension and mock them into place without fear of dislodging the housing from its position.

Below Left: The author recently modified the rear end of his Model A tudor chassis to give it more rear suspension clearance. The chassis is custom made from rectangular steel tube and uses a nine inch Ford rear end located by a triangulated unequal length four link system. The links are actually standard GM items from a Holden Torana. The coil spring mounts on the lower arm forward of the rear end and to the bracket on the outside of the chassis rail. Note the simple but strong shock absorber mounts. This is a simple suspension system that works very well.

Below: Few home workshops would have a jig for welding rear end housings but you can successfully carry out welding on your rear end housing by assembling it first. The axles and centre help to hold everything in alignment while you weld. Proceed carefully and slowly. Weld only in short bursts and allow time to cool regularly so that you don't put too much heat into the housing. Don't attach your ground clamp to the pinion or axle flanges or you will damage the gears and/or bearings.

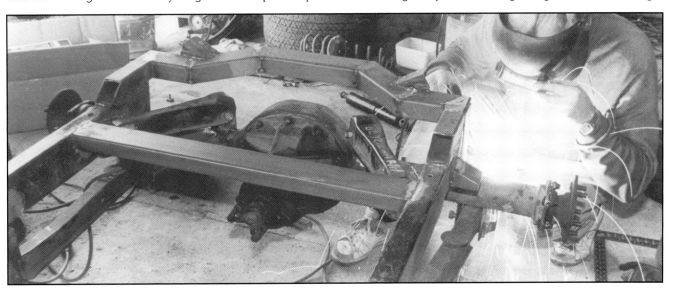

This four bar rear end system has forward mounts that are tied into the legs of the X member. A new crossmember is used at the rear to mount the coil-over shock absorbers. Note that the upper mount is adjustable thanks to triple mounting holes in the crossmember on each side. This will allow the suspension to be adjusted for best ride height and suspension movement once the vehicle is completed.

Slight stepping inward of the rear chassis rails allows large wheels and tires to be fitted. The inner fenders will also need to be modified to suit. For street use there isn't any real need to go bigger than this in the wheel/tire department. Note how the joint area features a large amount of overlap, effectively making the joint into a "box" and hence very strong. The rear legs of the tube center X member also tie into this joint area. A triangulated four link and coil-over suspension system will work well on this type of car. This is a very good example of how to set up this type of rear suspension. The rear end itself is a nine inch Ford unit.

four bars, but make sure your forward ends of the radius rods are mounted as close as possible to the center line of the chassis. This will allow the rear end to rotate about its axis with minimal bind. If you mount these early Ford type radius rods out at the chassis edges you will find your suspension is constantly working in a bind. Ride and handling will be adversely affected as a result.

Apart from setting the pinion angle at the correct point there is little involved in setting up semi-elliptic springs. Just make sure your springs are as long as you can get them without extending beyond the rear edge of the chassis and be sure they are parallel with each other and square to the chassis. The simplest way to achieve a good result is to duplicate the original factory set up that the springs came from as far as is possible. Again adding modern shock absorbers is highly recommended and all you really need is to provide upper and lower mounts, ensure that the shock absorbers aren't tilted on too much of an angle from the vertical and make sure they will clear everything around them during the course of their travel. Also make sure the suspension bottoms out on its bump stop rubbers before the shock absorber bottoms out in its housing or you will have bent shockers and broken mounting brackets in no time at all.

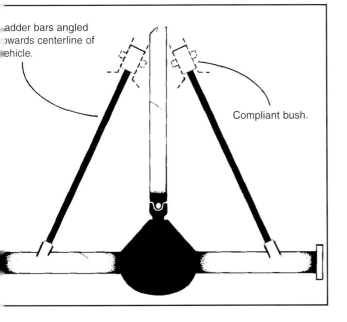

Ladder bars angled towards centerline of vehicle.

Compliant bush.

Left: *When ladder bars are used to locate the rear end in a street rod they should always be centrally mounted at the forward end so they rotate about the drive shaft. Don't fit straight ladder bars with their forward mounts out at the chassis rails for a street driven vehicle or the suspension will only be able to operate in one plane. This is dangerous on a street rod. The centrally located ladder bars essentially duplicate the method used on early Fords but with an open drive shaft.*

Below Left: *Worm's eye view of this '32 Ford chassis reveals a centralised ladder bar locating system that looks good and works well. The small tubular member that acts as the forward mount for the ladder bars also doubles as a drive shaft safety loop. Also note the well designed and fully captive lower coil-over shock absorber mounting.*

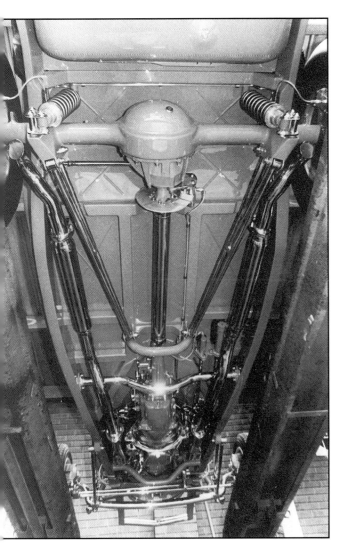

Below: *Close up view of this rear end set up shows how lots of adjustability has been built into the whole arrangement. Ride height and shock absorber angle is adjustable at top and bottom mounts and the panhard bar can be moved to a second chassis mount hole if required. Standing the coil-over more upright will stiffen up the suspension while laying it over more will soften the ride.*

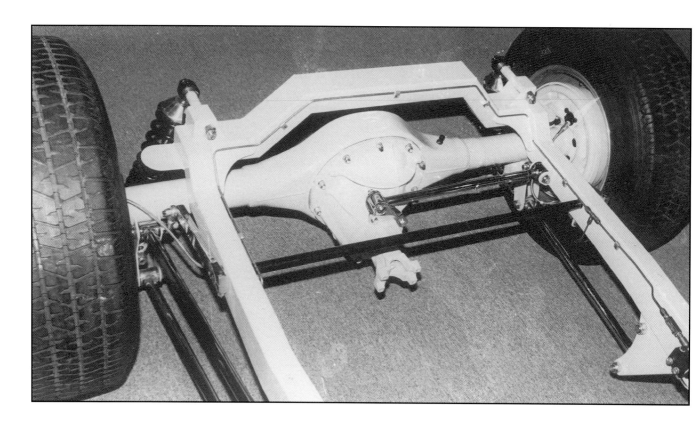

Tidy four bar and coil-over suspension on this Model A chassis will work well. Sway bar mounts through the chassis rails with splined arms at each end linking it to the rear end. Finally the panhard bar will keep it all centered in the chassis. This panhard bar is actually a little short, the longer you can make it the better.

Rear view of a four bar suspension system shows some really thoughtful engineering. The panhard bar is securely mounted at both the chassis and rear end and is level with the rear end housing. Note how the center X member also doubles as a drive shaft safety loop.

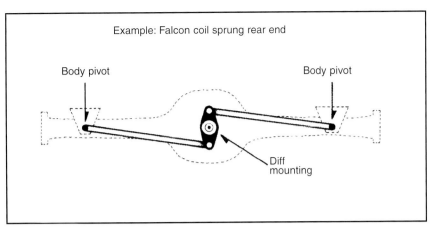

Example: Falcon coil sprung rear end
Body pivot — Body pivot — Diff mounting

An alternative to using a panhard bar for lateral location of the rear end is the Watts linkage. This system keeps the rear end perfectly centered in the chassis while it moves though its travel. The ends of the arms attach to the chassis while the central pivot attaches to the rear end.

Centralised ladder bar locating system on the author's own '32 hiboy chassis features a large rubber forward pivot joint from an English Range Rover. The Watts linkage mounted in front of the rear end where it is out of sight from rear view is a standard item from Australian Ford Falcons. Suspension movement is such that the cross links of the Watts linkage won't contact the ladder bars even though one link passes though the ladder bar.

Custom made watts linkage has uneven bars to locate the pivot point slightly to one side of the center. Existing bolts on the rear end housing and center are used to mount the pivot brackets. When the car is completed this type of locating system becomes almost invisible as it is largely hidden under the vehicle.

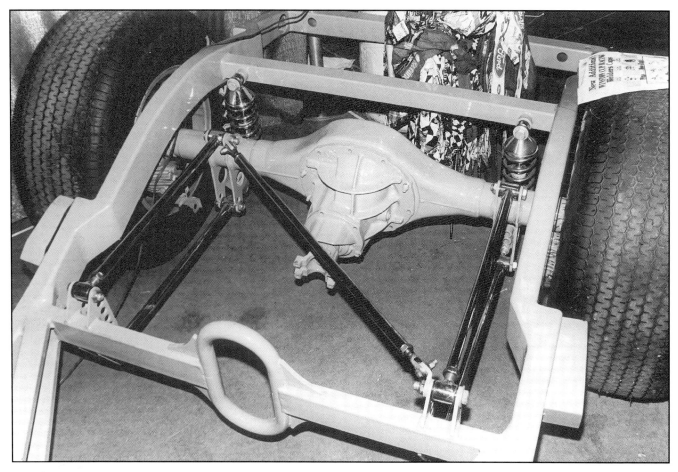

Stepped rear chassis rails shown here could do with more reinforcement at the splice. The four link locating system displays an alternative to the use of a panhard bar. The diagonal link performs the same function but needs to be carefully designed so as not to create binding in the suspension system. Note how forward and rear pivot points of the diagonal link align with the pivots of the four link arms.

Simple but efficient rear suspension systems can be engineered at home. The example shown here uses the spring and general layout from a Holden Torana but the locating arms are custom made. The top arms are in fact from a Nissan while the lower arms were made up from an old Ford wishbone with Nissan rubber bushes added at each end. Shock absorbers are from a Volkswagen and the rear end itself is eight inch Ford. It's a tidy installation that works well but didn't cost a fortune to manufacture.

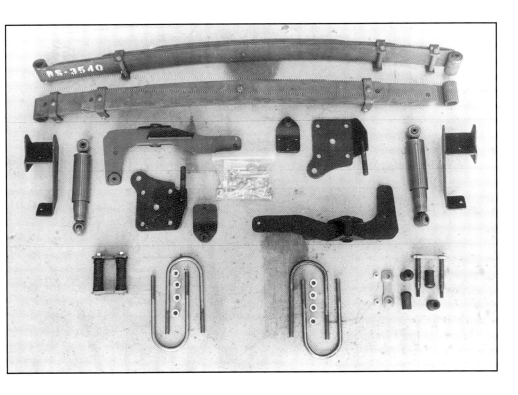

Here's a complete bolt-in semi-elliptic rear suspension kit for an early Ford chassis. Springs have reversed eyes at the forward end for maximum lowering and the chassis mounts are designed to bolt into existing bolt holes in the chassis where possible. Spring retaining plates incorporate the lower shock absorber mount and the sway bar mount.

Here's another complete bolt-in rear suspension system that would be ideal for the home based enthusiast who doesn't have well developed mechanical skills. The whole arrangement including the shock absorber crossmember bolts into the Chevy chassis.

Another simple semi-elliptic rear suspension set-up, this time in a Chevy chassis. The sway bar is mounted to the bottom of the rear end with a rod link to the chassis. Also note the forward mounting of the shock absorbers.

Basic semi-elliptic spring rear suspension system as fitted to a '32 Ford chassis. Note lowering block to give the rod a low stance but at the expense of having to take a 'C' section out of the chassis rail for additional clearance. This works quite well but be careful not to take too much out of the chassis rail or you may compromise its strength.

Simplest of all the rear suspension systems is the semi-elliptic parallel spring arrangement. Of particular note on this example is the slight sinking of the forward spring mount into the bottom of the chassis. This will help keep the car low to the ground without sacrificing ride quality. Clever engineering has seen the sway bar mount incorporated into the spring retaining plate. A simple crossmember provides the mounting point for the shock absorbers and the sway bar mounts. A simple but very effective installation.

Bolt-in semi-elliptic rear suspension systems make setting up this type of rear suspension very easy. This one includes a sway bar that bolts to the chassis immediately behind the crossmember where it will be almost completely hidden once the car is completed.

Quite large rear wheels and tires can be fitted to most fat fendered Fords just by straightening up the inner fender well as shown on this '35 sedan. Semi-elliptic springs locate a nine inch Ford rear end and taller than standard mounting blocks help lower the ride height.

It's hard to beat the fancy appearance of a well detailed Jaguar independent rear end installation in a street rod but mounting such a rear end is critical to its efficient operation. Study the diagram below for important mounting and alignment information for this type of suspension system when mounted without its original cage as is usually the case in a street rod.

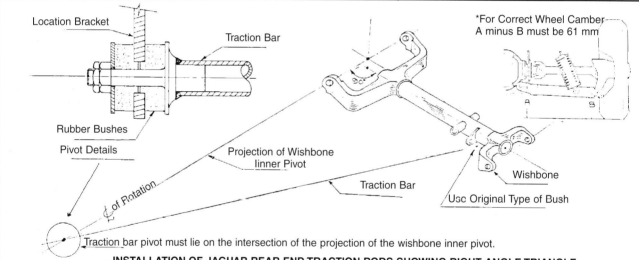

— INSTALLATION OF JAGUAR REAR END TRACTION RODS SHOWING RIGHT ANGLE TRIANGLE.

MOUNTING JAGUAR INDEPENDENT REAR SUSPENSION
RADIUS RODS:
Radius rods should always be installed to support the lower control arms of the Jaguar IRS.

There is only one geometrically correct position for them to be mounted. The radius rods must be installed with the forward end located on the axis of the lower control arm pivot. (The center line of the shaft that connects the lower control arm to the differential housing). This axis and the radius rod combine to make a right angle triangle with the lower control arm. This allows full suspension travel without bind. (See diagram). Mounting the forward end of the radius rod anywhere else other than on this axis line causes partial or total bind and undue stress on lower pivot bracket bolts which can break.

The recommended minimum size of tubing for radius rod fabrication is 1-1/8" O.D. The type of pivot used on radius rod ends should be of such design to allow adequate twist under full suspension travel, such as stock Jaguar rubbers on control arm ends and independent suspension strut rubbers on the forward end.

The other method of fitting radius rods is where the radius rods cannot be fitted in the same axis as the lower control axles, and the rods are mounted directly forward of the lower control arm connecting point, (as with stock Jaguar). If this type of system is used, the rubbers both for and aft should have sufficient movement to allow full travel without bind.

TIE BARS:
Tie bars or plates should be fitted between the lower control arm pivots; both front and back. Also a tie plate should be fitted underneath between pivot brackets.

TORQUE REACTION STRUTS:
These must be fitted between lower control axles and chassis to prevent the differential from twisting.

SHOCK ABSORBERS AND SPRINGS:
Mounting points for shocks should be the same dimensions as they were in the parent car. If this cannot be done for clearance reasons a minimum distance between top mounts would be 21". A correctly installed Jaguar IRS would have shock centers of 11-1/2" and horizontal half shafts. Car height under normal load, should be altered by changing springs, not relocating shock mounts. Chroming of springs is not recommmended, but if they are, they should be heat treated or sagging and/or breakage will result.

CAMBER:
Camber is adjusted by the use of shims between drive flange and half-shaft, and bottom pivot bracket and differential case. Correct camber is 3/4° plus or minus 1/4° negative.

WHEEL BEARINGS:
It is very important to adjust bearings correctly. This is done with varying size shims to accomplish an end float of .002" to .006". These bearings are not preloaded. If they are, severe damage to the hubs will result.

Above:
Jaguar rear end in a Chevy chassis attempts, to some degree, to duplicate the original Jaguar mounting in that the centre section is mounted on insulated bushes. This will help minimise noise transmission into the chassis. Strut rods at the front control the rear end's natural tendency to "wind up" under hard acceleration.

Above Left:
This Jaguar independent rear end installation in a '33 Ford shows how the forward ends of the outer radius rods should be mounted so that they align with the lower pivot points of the rear end. This allows them to operate without binding. Don't mount them straight along the chassis.

Left:
Another independent rear end but this time a Corvette unit mounted in an early Ford chassis. The original Corvette mounting frame is used at the rear with mounts to attach it to the chassis. Forward crossmember holds the pinion of the rear end in position and prevents wind up.

CHAPTER 5
MOUNTING THE DRIVELINE

Study the rubber mounts and their original brackets in a factory car and then work out a way to incorporate those design elements into your own engine mounts.

ENGINE MOUNTS AND CROSSMEMBERS

There's one simple rule to follow when mounting your engine, where possible incorporate the original factory rubber mounts if you can. They were designed to hold that engine in the most efficient manner so you may as well take advantage of the factory engineering research. Be careful with the design of brackets to hold or locate the rubber mounts too. Your mounting hardware should be designed so that if the rubber mount fails, the engine cannot drop onto the roadway or foul your steering components. Study the rubber mounts and their original brackets in a factory car and you will see how this has been done. Then work out a way to incorporate those design elements into your own brackets.

Many engine brackets are designed so that they support the weight of the engine on the bracket itself, not its mounting bolt, so that the rubber mount is really acting as an insulator. Chevrolet V8 engines are a typical example. If you study them carefully you will note that there is a flat surface incorporated into the mounting bracket that locates against the back of the rubber insulator. This flat surface supports the weight of the engine and the bolt that passes through the mount is only there to hold it in place on the bracket. The bolt itself should not support the weight of the engine. Be observant and you will find that most engine mount insulators use the same theory in some form or other.

Wherever possible you should incorporate an engine mount crossmember as well. This should extend from chassis rail to chassis rail with brackets attached to accept the rubber insulators. Make it a bolt-in fitting if you can to make maintenance easy at a later stage. Using a crossmember like this will overcome the tendency of the engine weight to try to spread or rotate the chassis rails. It will also serve as an extra safety device if you should ever have an engine mount failure. If the engine crossmember passes under the engine it will serve as a restraint that could prevent the engine from dropping to the roadway or onto your steering components.

One further benefit from using an engine

This engine mount crossmember is to suit a small block Chevy engine. The flat plates on top of the locating tubes support the weight of the engine in the same manner as the original Chevy item. The bolt though the tube is there to hold the mount in place only. Don't rely on a bolt through the mount without the flat plate on this type of engine mount. The bolt should not support the weight of the engine.

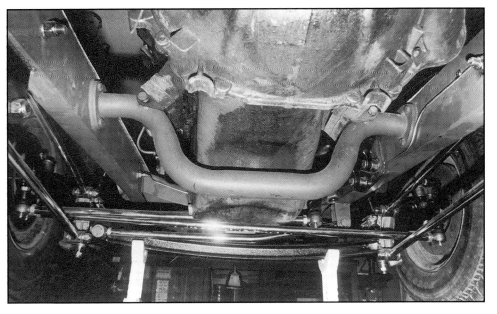

There are several advantages to using an engine mounting crossmember like this one. It helps tie the chassis rails together but is removable for later maintenance. If you should decide to swap in different running gear at a later stage you won't have to remove the engine mounting brackets from the chassis rails either, just make a new crossmember to suit. In the event of an engine mount failure the crossmember will also prevent the engine from dropping onto the steering or the roadway.

mounting crossmember to secure your engine is that any future change of engine will only require the manufacture of a new crossmember. You won't have to cut away old engine mounts or do any chassis or paint repairs to fit your new running gear.

Should you be updating an existing street rod in the driveline department you can still add a margin of safety to an existing engine mounting arrangement if it doesn't incorporate a crossmember. Some engine mounts consist of simple gussetted brackets welded to the chassis rails and that rely on the mounting bolt to support the engine weight. Often a support plate can be added to such a system as described above and it is a good idea to at least add a simple bolt-in tie-bar to link the separate brackets together under the engine. If nothing else your peace of mind will be enhanced by these additions.

TRANSMISSION CROSSMEMBERS AND MOUNTS

Just as it's a good idea to use a crossmember under your engine, so it is under your transmission. Make it a bolt-in item and again your future maintenance access will be considerably enhanced. Use the factory rubber insulator mount for your transmission if possible and ensure that it isn't in a bind once fitted. If it is under tension when installed it will fail rapidly.

The bracket that accepts the rubber insulator needs to be carefully designed. It needs to be properly gussetted to withstand the weight of the driveline but it also needs to adequately clear the transmission itself in the vicinity of the mount. Even if only slightly touching the extension housing of the transmission the resulting vibration will be very annoying for the passengers inside the vehicle.

Take the time to think about the design of your transmission mount well in advance of actually making it. There are several things to keep in mind

A complete bolt-in transmission crossmember will allow for easy maintenance later and if you should change the running gear at a later date you will only have to make a new crossmember. In this instance the transmission crossmember incorporates the mount for the brake master cylinder/booster as well.

as you design the mount. Remember to allow clearance for the speedo cable drive and for any shifter linkages that could cause interference in use. Also consider that you may need to locate exhaust components in this area so remember to allow space for them early in the designing process.

Depending on the particular shape and location of your transmission you may find that you can incorporate a driveshaft safety loop into your transmission mount too. There's more on this subject coming up later in this chapter.

SETTING THE ENGINE AT THE CORRECT ANGLE IN THE CHASSIS

Before you get into the mounting of your engine and running gear you will need to have established the rake of the finished vehicle. With the chassis set up at this angle you can begin the process of locating the engine/trans package. The starting point should be the carburetor base of your intake manifold. When the engine is installed in the chassis this should be as

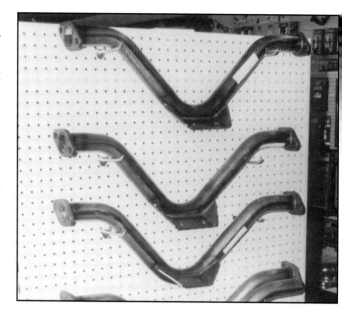

Complete bolt-in transmission crossmembers are available for common driveline swaps into early chassis. As with engine mounting crossmembers this type of mount allows for easy driveline changes later in the life of your street rod. Being completely removable also makes for uncomplicated removal if you should need to remove your transmission for maintenance.

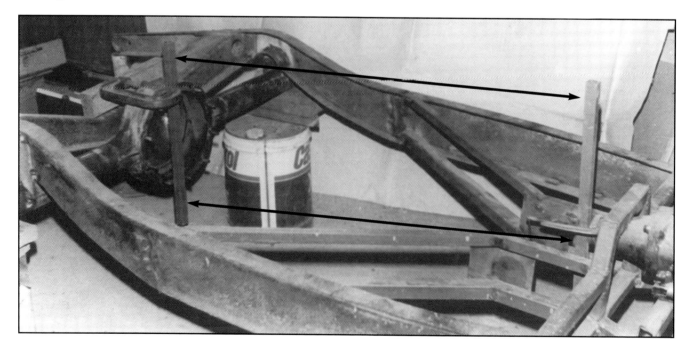

For correct driveshaft universal alignment the pinion flange and the rear face of the transmission extension housing should be at the same angle. However some like to have the pinion angle one or two degrees down to allow for wind up under power. One way to set this angle is to clamp straight edges to the front face of the rear end (or the pinion flange if your diff center is in place) and the end of the extension housing. Measure between the straight edges at top and bottom and make sure the measurements are the same. Now your universals will be aligned to each other when the driveshaft is fitted.

close to parallel with the road surface as possible. Usually this will result in the engine being at or about three degrees from horizontal.

With the transmission attached to the engine the rear of the transmission extension housing should now be aligned with the pinion of the rear end. The transmission doesn't have to be directly lined up with the pinion in a straight line but there is some important geometry to take into account at this point.

A straight edge placed across the vertical surface of the pinion flange on the rear end will give you a reference point for mounting the rear of the transmission. A straight edge placed vertically across the end of the Transmission extension housing should be in alignment with the one at the pinion flange. This will ensure that both the front and rear driveshaft universals are working in the same plane. If they aren't, your driveline will destroy universals on a regular basis, so take the time to get these angles matching each other. Usually the driveshaft will be centered when viewed from above but not always. Even if yours is offset to one side, which is sometimes necessary for clearance in a tight engine bay, the same alignment process needs to be followed horizontally as you did for vertical alignment. That is the angle at the pinion flange needs to be the same as the rear of the transmission in a horizontal plane. If you have your rear end centered properly in the chassis and your engine/trans package centered in the chassis your horizontal driveshaft alignment should automatically be correct. Don't take it for granted though, double check it just to be sure.

Despite the information just given there are some engineers who advocate offsetting the engine/trans combination on a very slight angle to make the universals "work" more in operation. The idea behind this concept is to prevent the universals from wearing in the one spot as a result of being perfectly aligned My own opinion is that there is so much going on down that driveline when the vehicle is being driven that either method will prove satisfactory in most cases.

Once you have established all these angles and have the engine/trans package located in the correct position in the chassis you can go ahead and make the mounts for engine and transmission as described earlier in this chapter.

DRIVESHAFT SAFETY LOOPS

In this age of high safety consciousness it is amazing that new vehicles with rear drive are still made without driveshaft safety loops. Even with everything set up correctly in the driveline department there is still the risk that a universal could fail from faulty manufacture or lack of maintenance. If it happens to be the front universal there is a strong risk that the driveshaft will drop to the roadway while the vehicle

A simple U shaped piece of steel tube has been incorporated into this Model A Ford chassis to act as a driveshaft safety loop. The brake pipe has been positioned such that in the event of a driveline universal failure the driveshaft can't come into contact with it. Always try to incorporate as many safety related items like this as you can into your street rod. The peace of mind is worth it!

Left: *Clean rectangular tubing center X member in this chassis has been made in such a way that the rear section also acts as a driveshaft safety loop. If the front universal fails there is no way the driveshaft can drop to the roadway and even if it is flailing about it won't come into contact with any other mechanical components.*

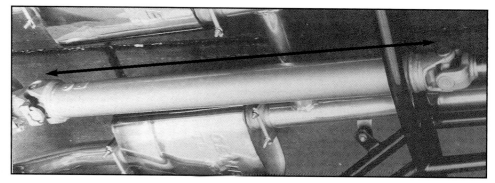

Right: *Just as the universals in a steering system need to be mounted "in phase" so do they in a driveshaft. The weld yokes in each end of the driveshaft must be aligned to each other or you will have a driveshaft that vibrates and destroys universal with monotonous regularity.*

is at high speed. If the driveshaft digs into the road surface in such a circumstance the resulting accident will be very severe. Don't take that risk, it is an easy thing to include as you build your car!

All that is required to overcome such a dangerous situation is to make some sort of retaining device that will "catch" the driveshaft in the event of a front universal failure. This could be as simple as an extra crossmember with a dip in the middle to keep the driveshaft from escaping from between the member and the floor of the vehicle or better still it could be a complete loop of some design that will totally capture the driveshaft and prevent it from coming into contact with the rest of the undercarriage. Go for the second option if you can, it will also ensure that less damage is sustained by the undercarriage if you do have a universal failure.

DRIVESHAFT PHASING

There is one more important factor to take into consideration when it comes to making and fitting a driveshaft in your street rod. Take a look at any factory made driveshaft and you will see that the universals at both ends of the shaft are aligned to each other. That is the yokes that are welded into each end of the shaft are aligned to each other along the shaft. This is called driveline phasing. If the yokes aren't aligned like this you will definitely experience driveline vibrations that will lead to eventual failure of the universals. If you need to have a driveshaft custom made to suit your street rod make sure it is done by a competent machine shop that understands driveline phasing.

CHAPTER 6
BRAKES

Probably the single most important part of your street rod is the braking system. Modern traffic moves at a fast pace and you need to be able to match it in all aspects of performance, handling and braking.

Probably the single most important part of your street rod is the braking system. Modern traffic moves at a fast pace and you need to be able to match it in all aspects of performance, handling and braking. Very few modern cars are made with drum brakes and there is a reason for that. They are not as efficient as disc brakes. If you are building a nostalgia style car you may wish to retain drum brakes for originality but my personal preference would still be to use discs. Safety should always come first.

There are several specialist street rod braking system suppliers now in the marketplace and I urge you to use their expertise if possible. If this isn't possible or practical at least try to make your brake system as compatible as possible. Use all the components from a single source if you can and ensure that the donor vehicle is at least as heavy as your rod.

The problem of selecting an appropriate master cylinder or master cylinder/booster combination for your street rod is not always straight forward but we will attempt to make the selection a little better informed. One fundamental rule is that the disc or drum brake used determines the master cylinder and booster selection and that the discs or drums must be chosen for their suitability to your particular car. It is not possible to make an inadequate brake design adequate by changing the master cylinder or booster, which gives us our first guidelines.

It is hard to imagine a tidier brake system than this one complete with finned caliper mount and dust plate. Note the smooth radii on the hard lines and the through the frame fitting to the flexible hose. In most jurisdictions the braided flexible hose shown here would not be acceptable as it often won't have a DOT rating.

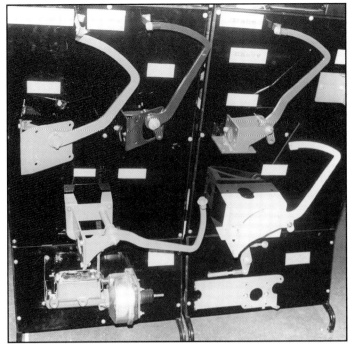

Now that we have brakes on both ends of the car, we can choose the master cylinder and booster. Basically you must choose components equal to those fitted to the car from which the foundation brakes came, unless your car is significantly lighter than the original brake source vehicle, which would permit a reduction in brake line pressure but not a reduction of master cylinder displacement, which brings us to the next point.

3. Master cylinder displacement (area of bore x stroke) should never be reduced over that of the car from which the brakes came. When your brakes are installed, a normal stop should never use more than half the total master cylinder piston travel. A further check should be made when a tandem

Above: *It is important to have a pedal mechanism that works efficiently. Fortunately if you can't adapt an existing unit or make one yourself there is now a wide selection available from manufacturers such as Butch's Rod Shop. Many of these are a direct replacement for the tired or unsuitable original in your early car.*

Right: *Another advantage of using one of these ready made replacement pedal mechanisms is that they are usually made to accept a modern master cylinder and servo system.*

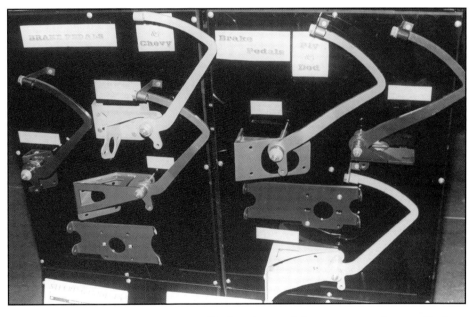

1. The foundation brakes (i.e. wheel brakes) for your car should be at least as good as those off the car from which the engine came. e.g. If you are using a 351 Ford engine, use the ventilated Ford brakes from the donor vehicle if possible. Remember that on any street rod, the front brakes will do more stopping than the rear. Therefore, the front brakes should always be bigger and have a greater torque capacity than the rear. This determines our second rule:

2. Having established the front brake selection, the rear brake disc, drum wheel cylinder or caliper piston diameter should not be greater than the same part on the front if you are to be able to balance the brakes easily.

master cylinder is used because tandem cylinders usually have different displacements between the two halves. The half with the greater displacement will connect to the front brakes which of course will have the greater piston area. To ensure that there is adequate individual displacement from both halves of the master cylinder, the following test should be used.

Apply the brakes hard with a single application. While the brakes are applied have a helper release a front bleed screw. Note the remaining pedal height and then release a rear bleed screw. If there is no additional pedal movement on the last bleed screw opening, then there is insufficient rear brake

A little forethought when making your center X member will allow you to incorporate your pedal and master cylinder mount at the same time. Use as large a vacuum booster as you can fit in the space available and try to use factory matched components wherever possible.

This pedal and master cylinder/booster system fits well in the available space but beware using a booster that is too small. Such a small booster in a large car will probably work okay for one or two panic stops but reserves will be limited in extreme situations or when towing a trailer.

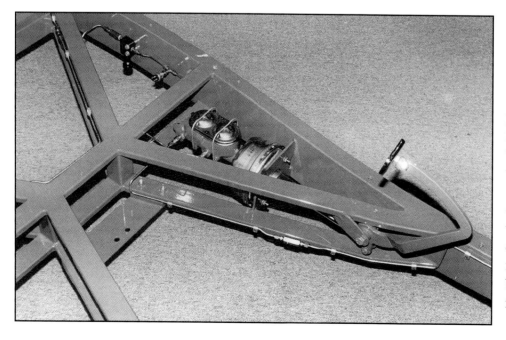

displacement and the rear brakes will lose effectiveness. The same test but releasing the rear brakes first and then the front would indicate if the master cylinder was connected incorrectly.

NOTE: This assumes that the full master cylinder stroke is available. Both master cylinder pistons must be able to fully bottom if a tandem master cylinder is to work correctly. This is usually about 1.4 inches input push rod movement. If this test indicates insufficient displacement ensure that the brakes are correctly bled and adjusted. If they are, there is either a malfunction in the system or you need a bigger master cylinder.

BOOSTER SELECTION

You would be unlikely to be wrong if you selected a booster off the car that the wheel brakes and master cylinder came from, unless your car is significantly heavier than the original. However, if you intend to design your own hybrid system as hot rodders are inclined to do, you need to make some good judgements. There are in common use, two types of booster systems. A vacuum mechanical booster which fits between the pedal and master cylinder known as a MASTERVAC and a vacuum hydraulic booster to fit in a hydraulic line known as a HYDROPOWER.

MASTERVAC

The function of both types is the same. Their purpose is to increase hydraulic pressure. In the case of a mastervac, the output of the unit is determined by the area of the diaphragm (some units have two diaphragms) multiplied by the pressure differential available (vacuum), the rate of boost application being controlled by a valve in the input pushrod

Left: *Master-vacs are the units that operate directly on the master cylinder. The larger version on the right has two diaphragms inside giving greater output than the single diaphragm unit at the left. If you have room in your street rod go for the double diaphragm unit.*

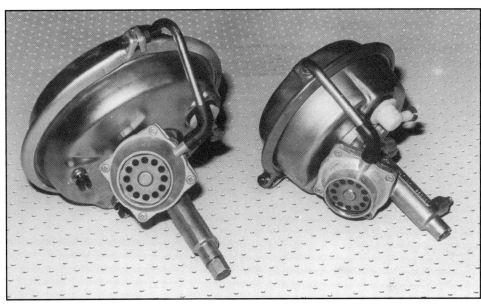

Right: *Two popular sizes of hydropower boosters. The unit on the left has a larger diaphragm for applications where high pressure and high displacement are important (see text).*

Below: *Here's a typical commercially available pedal and master cylinder mounting system as fitted to an early chassis. A pedal shaped like this one will only require a relatively small hole in the floor.*

assembly. Too small a unit will give a limited output and will result in a strange pedal feel with the brakes feeling okay on initial application but developing a dead feel as the boost runs out. By the way, that bottoming out condition you feel with mastervacs when you stand on the pedal is not the end of the stroke, but the full opening of the control valve and the point at which the boost will run out. You should not experience this run out condition when the car is being driven normally.

HYDROPOWER

Hydropowers are a bit more complex as their output is determined by the area of the diaphragm multiplied by the pressure differential and the area of the slave cylinder piston which also limits the displacement of the unit (area of bore x stroke). The following

BALANCING BRAKE SYSTEMS

Fine balance is one of those concepts that needs to be applied to most things in life and hot rodders in particular will understand this need especially as it applies to mechanical devices like engines and things that revolve at high speed. So it is with a braking system. Despite the fact that you may have on either end of the car the best brake system that money can buy, if they are not balanced to each other, the brakes will be ineffective if not dangerous. The brake performance of any vehicle is limited by the locking of any wheel. If for instance a rear wheel locks and at that time the front wheels are only doing 50% of the braking that they are capable of before front wheel lock would occur, then the available deceleration is drastically reduced to just a little over half of what would otherwise be available. With any modified car, assuming that the basic design is correct, the brake system must be balanced for even reasonable performance. The object of this exercise is to get both ends of the brake system doing exactly the amount of braking able to be applied without premature wheel

Above: *Careful selection of components and some clever bracket fabrication has even left enough room to incorporate a hydraulic clutch system in this '34 Ford chassis. Pedals will operate either side of the steering column where space is tight but adequate. Always use a dual circuit brake system on your street rod.*

Above: *In-line type proportioning valve is adjusted by fitting a different length spring. The unit needs to be disassembled to change the spring necessitating rebleeding the system on completion.*

guidelines should be applied. Never use a hydropower with less displacement than the master cylinder and choose a unit which has a boost run out pressure that is higher than the normal brake line pressure requirements of the car, none of which is easy to determine. To be sure you have the right one, consult a brake specialist.

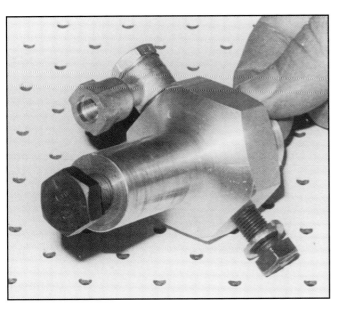

A screw adjustable proportioning valve like the one shown here can be reset without the need to disassemble any of the brake system. See the text for details of testing for correct brake balance.

TESTING FOR BRAKE BALANCE

The test for properly balanced brakes is quite simple:
1. Load the car as it would normally be in operation.
2. Run the car on a good surface at say 60 MPH, and apply the brakes slowly until wheel lock occurs. Note which wheels lock first.
3. The braking effect of the axle that locks first must be reduced or the axle that doesn't lock must be increased until under these conditions (with the brakes relatively cool) both front and rear wheels will lock at the same time or very close to it. Generally, most cars have a rear brake bias to allow for a load increase on the rear axle and it is usual to reduce the rear brake performance to improve the balance. DO NOT FIDDLE WITH LINING AREAS as this has no effect on the friction produced by the brakes, but can upset the geometry of the design. If you get a result by this method it will not be for the reasons you believe. What you must do is to change the forces applied at the end of the brake shoe or behind the brake pad. This can be done by altering the piston size or the line pressure.

lock. This of course is a function of the weight on each wheel and the friction available between the tire and the road. The difficulty is that this changes as a car becomes laden and as the rate of deceleration increases. The higher the deceleration the higher the transfer of weight to the front wheels from the rear wheels.

Here's an adjustable proportioning valve as fitted to the author's Model A Ford pickup. Such a valve should always be fitted into the rear brake circuit in a position where it is reasonably accessible for adjusting purposes but where it is secure from risk of damage. There are several different designs of screw adjustable proportioning valves but all essentially work in the same manner. Choose the one that best suits your needs and available mounting space.

METHODS FOR ALTERING BALANCE

All things considered it is usually easiest to change the pressure to the rear brakes with a proportioning valve. This has the added benefit of putting a high share of brake load on the rear brakes during low decel stops where wheel lock is not a problem. This evens up wear between front and rear brakes as most stops are from relatively low speeds at low rates of deceleration.

Two types of in-line proportioning valves are readily available, one being adjustable by changing springs, the other by turning a screw. The spring adjustable design is also incorporated in some late model master cylinders which to adjust, the valve or cylinder must be removed and a spring changed to gain either a higher or lower pressure. The screw adjustable type is able to change the pressure (always a greater or smaller pressure reduction in either case) at the turn of a screw, which greatly simplifies the process and enables you to alter the brake balance while you drive. If you are to be safe driving in heavy traffic in your street rod you must be able to stop at a rate similar to other cars and with all other things being equal, it's just a matter of balance.

PIPING UP HYDRAULIC BRAKE SYSTEMS

Just as arteries are to the body, so are the hydraulic lines to your brake systems. There is nothing quite so exciting as a total brake failure and about the only way it can occur is by a pipe or hose failure. Of the hydraulic brake line failures that do occur, most happen due to faulty or careless installation.

PIPELINE LAYOUT

Where possible run the pipeline on the inside of the chassis, at the rear of the axle and above any crossmember if possible to minimise the chance of stone damage. It is also imperative to keep hoses and pipes well away from any exhaust pipe and even then, if in doubt fit a heat shield.

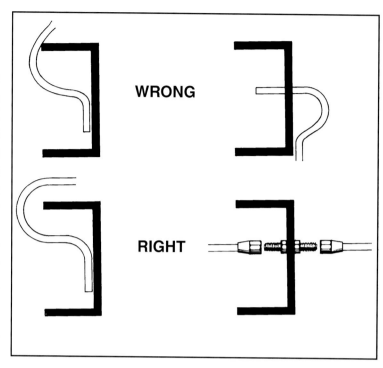

Great care must be taken when routing hard brake lines around your chassis. All bends should have a smooth radius with no kinks and the hard lines should not contact any sharp edges. Where the hard lines need to pass through a chassis rail or crossmember it is usually better to use a junction fitting as shown. Make sure such fittings are proper brake components with the correct flares.

If the body of your street rod is not rigidly mounted to the chassis and your master cylinder is firewall mounted, consider using a hose to join between the two. Pipe breakage on some off-road vehicles for this reason is quite common and it can occur without warning.

PIPE MATERIAL

Plain old steel tube is probably the best you can use. It is readily available, easy to bend and flare and if fitted properly and kept clean will never wear out. However hot rodders being what they are will want something different so you can use stainless steel tube if desired. It bends and flares okay and will never wear out, but it is harder to obtain and it tends to work harden. For this reason some licensing authorities will not allow its use. Check first! It has been common in the past to use heavy wall copper pipe for brake lines but this is now advised against and is illegal in some jurisdictions.

Most automotive brake pipes are 3/16" diameter. Some older cars used 1/4" and metric pipe sizes while not being identical are usually interchangeable. If the port sizes are 3/8" or 10mm use 3/16" pipe. If the port sizes are 7/16 or larger it may be necessary to use 1/4" or larger pipe to pass the volume of fluid required such as on a truck or on a clutch line.

AVOID SHARP BENDS

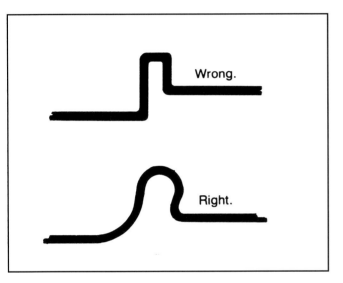

It is possible to trap air in brake lines by having too sharp a bend radius as shown here. If such a bend is necessary use larger radius bends to avoid air entrapment and pipe kinking. If possible avoid at all having upward loops in the pipe as these can make bleeding difficult.

TYPES OF FLARING TOOLS

There are cheap slow tools available from some equipment suppliers which can be suitable but are really only for emergency use. Good flaring tools are expensive but are essential if good work is to be done in any volume. These types of tools are often a good investment for clubs because they are only used occasionally.

TYPES OF FLARES

There are now basically three types of flares in popular use.

1. SAE Double Flare

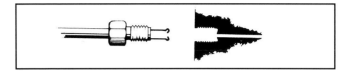

This is the most common type found on American origin vehicles. It is always used with an inverted seat usually with 3/8" -24 UNF or 7/16" -24 UNF thread tube nuts on the brake pipe.

2. SAE Ball Flare

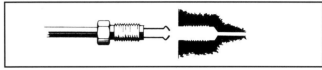

This is used on a drill point seat and is the first stage of a double flare operation. The tube nut is always longer than a double flare nut and usually with 3/4" -24 UNF or 7/16" -24 UNF threads.

3. ISO Flares

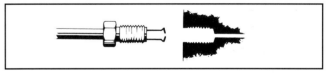

Similar to the ball flare but with a flat seat shape as opposed to the tapered ball shape. Usually with M10x1 metric threads.

Japanese flare systems are often a mixture of all three with metric threads.

PIPE CLAMPS

Many different types are available from exotic milled aluminum versions to the simple rubber cushioned ADEL clips. Whatever you elect to use make sure the pipes are secured every 12 inches or less to prevent the pipes from vibrating and contacting chassis or suspension components that might cause damage.

MAKING TEMPLATES

Where a pipe takes several bends in a short distance make a template from welding wire and then bend

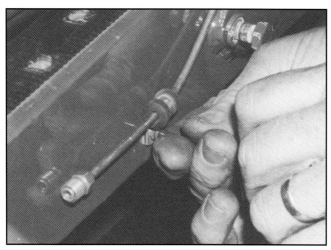

Above: Rubber cushioned Adel type clamps are a good option for securing your brake pipes to chassis rails and crossmembers.

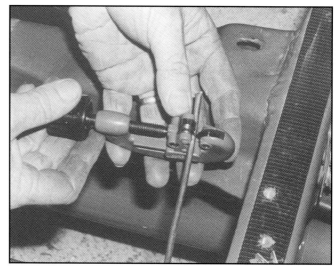

Above: A small pipe cutter will give you nice clean cuts that require very little cleaning up before making your flare on the end. This is much more satisfactory than using a hacksaw to cut your pipes.

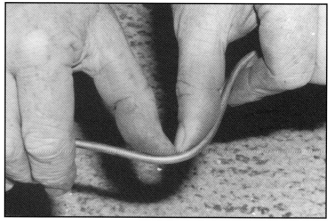

Above: Common 3/16" steel pipe will bend easily over your fingers but take care not to flatten or kink it in the process.

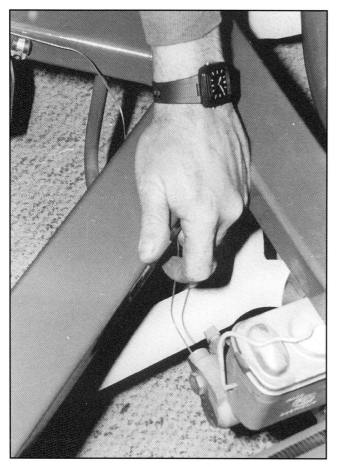

Above: To save frustration use welding wire to make a template for your brake pipes and then bend the pipe to match your template

the pipe to the shape of the template to save frustration. If a very short pipe is necessary make a template using welding wire, avoiding sharp bends, to establish how the pipe will look and to find the correct length. Memorise the shape, straighten the template and cut the pipe to length, remembering to allow extra length for the flares. Then bend and adjust the pipe to the final shape.

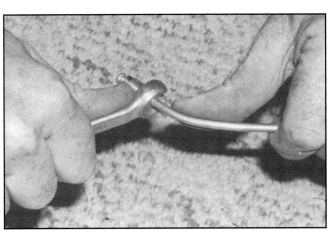

Above: *If a tight bend is needed close to the end of a pipe use a pipe nut wrench on the nut and your thumb as a mandrel*

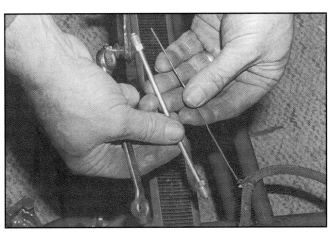

Above & below: *Where a very short pipe is necessary make a welding wire template, avoiding sharp bends, establish how the pipe will look and to find the length required. Memorise the shape, straighten the template and cut the pipe to length, remembering to allow enough extra length for the flares. Then bend and adjust the pipe to the final shape.*

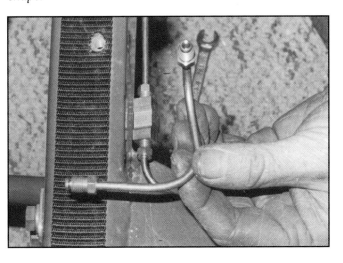

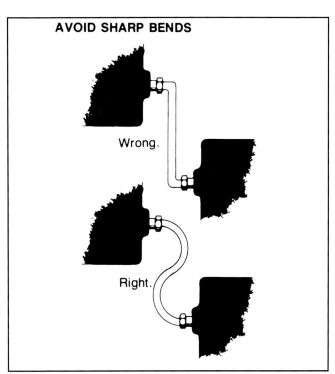

Above: *It is very difficult to get all the bends in the right place first time so you will need to adjust them.*

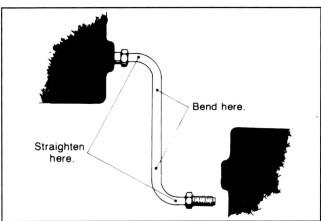

Above: *Provided you haven't made the bends too tight it is easy to adjust them as shown. Bends made on a mandrel are usually too tight and tend to work harden the pipe more. For this reason you may find it easier to form your bends by hand.*

SEAT AND CHECK FLARES

Having finished your pipe tighten it down firm but do not overtighten, then undo it and check the flare seat to ensure proper seating. If it is not seating fully refit and retighten. If still no good discard the pipe

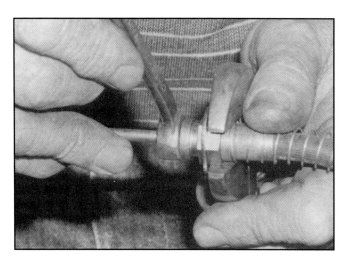

Left: *To seat a flare use the steel end of a brake hose rather than a brass union or aluminum caliper which may be damaged by a flare that won't seat properly. Tighten the nut down firmly but do not overtighten, undo it and check for proper seating.*

and make a new one. When seating a flare like this use the steel end of a brake hose, not a brass union or aluminum caliper which may be damaged by a flare that won't seat properly. It may even pay to keep an old brake hose somewhere in your workshop expressly for this purpose.

Above: *Route brake pipes inside chassis rails and crossmembers where they will be protected from stone damage.*

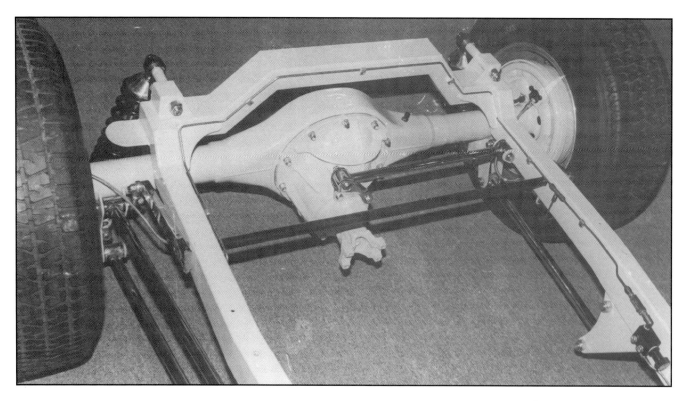

Above: *We showed you this photo in the rear suspension chapter but this time we'll concentrate on the brake lines. Note how the hard lines are routed inside the chassis rails and secured every 12 inches or so along the way. Routing the rear lines to an outlet on each side of the chassis and using flexible hoses to the rear bakes overcomes the need to run hard lines across the rear end itself. A tidier installation is the result. Note the adjustable proportioning valve mounted inside the chassis rail at lower right of photo.*

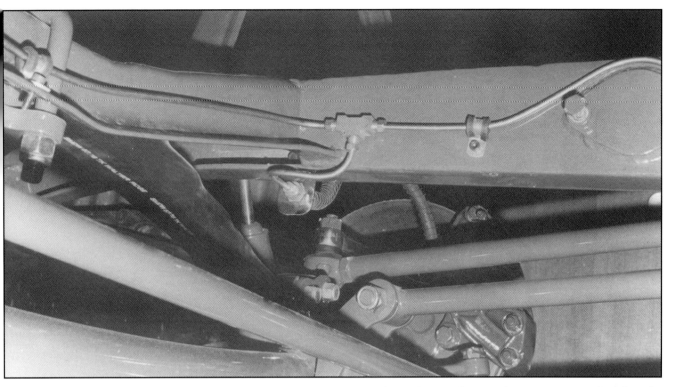

Above: *Where front feed pipes need to pass under the chassis rail as shown here make sure they are positioned so that the suspension or steering components can't come into contact with them during operation. Remember to allow for the fact that the steering tie rod moves forward toward the spring when the wheels are turned.*

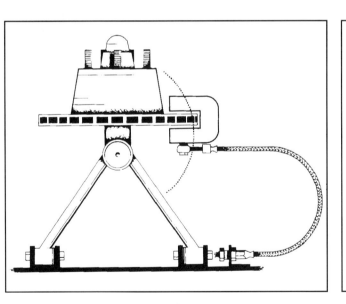 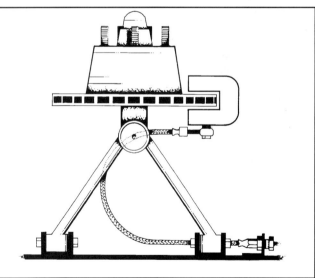

Flexible brake hoses should not be mounted in the manner shown at left. The hose will have to be longer than necessary to allow for the movement of the wheel through its turning arc. When the wheel is turned there is a chance the line could be stretched if it isn't long enough and the extra length also means it will be more likely to interfere with other suspension components. Flexible brake hoses should be routed as shown at right so that they pass through, or close to the ball joint or king pin pivot point. This minimises the amount of travel the hose needs to make when the wheel is turned through its arc. It also means the hose can be shorter than the first alternative.

Above: *Careful positioning of the flexible brake hose on this front end allows the hose to be quite short. Note how the hose passes through the pivot axis of the balljoints which helps minimise the amount of movement required when the wheel is turned from lock to lock. The front end is a Holden Torana that has been fitted with larger ventilated disc brakes and is mounted in a '48 Anglia chassis.*

Above: *The author updated the brakes on his Model A tudor from the narrow solid rotor shown on the left which is from a late sixties GM product to the larger vented unit on the right. Additional braking capacity is immediately obvious. Careful selection of components resulted in a track width that was also slightly narrower than previous thanks to the shorter hat section on the vented rotor. The vented rotor is from an Australian made Leyland P76 which features the same bolt pattern as popular Fords but will also accept a GM caliper.*

Left: *After making a brake change or adaption always check for adequate clearance in all directions. Turn the wheel through its complete steering arc and move the suspension up and down through its complete travel. Doing this revealed a minor problem on this installation where the caliper just came into contact with the upper arm of the suspension where indicated by the screwdriver blade. Just a little work with a hand grinder was required to provide the necessary clearance.*

If you overtighten to obtain a proper seat you may either swell the end of the nut which will strip out the thread upon removal or simply strip the thread from the tapped hole, particularly in aluminum. Either way it is not worth the hassle, make a new pipe or get a better flaring tool. Fully polished aluminum calipers can be expensive to replace so don't take the chance of ruining one by using it to seat flares.

Milled aluminum clips have been used here to secure the brake pipes to the chassis rail and front crossmember. The front junction block has been positioned so that when the front pipe passes under the chassis rail to feed the left front brake it does so just to the rear of the tie rod. This ensures that the tie rod won't come into contact with the pipe at full suspension travel. Note also how the pipe has been formed to clear the spring retaining U bolts to prevent vibration damage.

Mounting the hard lines across the front of the rear end as shown here makes the underside much tidier when viewed from under the rear when the car is finished. The extra tube fitted to the brake junction block is actually a breather for the rear end. The mounting bolt is hollow allowing the vapor to pass through it into the tube. The upper end of the vent tube feeds into the rear crossmember.

FITTING A PARKING BRAKE AND MOUNTING THE CABLES

Essentially there are two options when it comes to fitting a parking brake mechanism, you can buy new aftermarket components and build your own system from there, or you can make your own custom setup using parts from late model donor vehicles. Of course in some instances you may even retain the original arrangement in which case all that would normally be required is to rejuvenate the worn components.

Several suppliers can provide you with all the aftermarket parts you require to build a parking brake system from scratch. There isn't really much to it, just find a place to mount the handle mechanism

Simple tabs welded to the forward leg of the center X member provide a neat mounting point for a "through the floor" parking brake. As with many modern mechanisms like this a short length of cable extends from the handle to a junction block which also accepts the individual cables that pass to the rear end. The anchor point for the rear cables is a simple plate bracket welded to the center crossmember. The end result is a tidy installation that will work well and has plenty of adjustment.

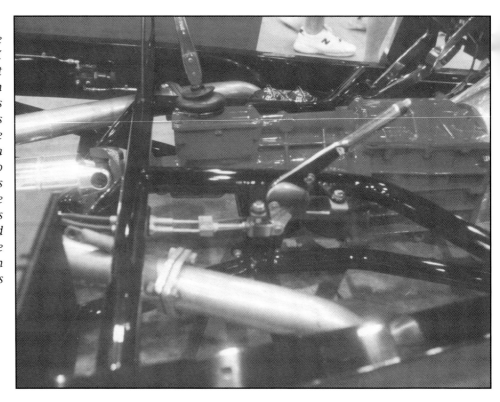

Many modern small cars have parking brake mechanisms like that shown below. These consist of a handle with a length of flexible primary cable attached. This makes it easy to mount the handle in a convenient location and route the flexible cable to the underside of the floor where it can be connected to the main operating cables. (See next photo).

A simple bracket anchors the rear end of the flexible primary cable in this parking brake installation. Another simple steel bracket picks up the ends of the operating cables and is locked with a nut each side after adjustment is completed. This entire system is made up from standard factory pieces apart from the two steel brackets.

that you prefer the look of the pinion mounted type because it is located centrally under the car and overcomes the need to have cables extending to the backing plates. One thing to keep in mind with pinion mounted parking brakes is their method of operation. Some are hydraulically operated and in most jurisdictions they are illegal. Generally the parking brake must be mechanically operated. Such versions of the pinion mounted parking brake mechanism are available and an example is shown in this chapter.

Pinion mounted park brakes are usually required to be mechanically operated for legal compliance. Advantages are single cable operation and no cables hanging out of your backing plates but extra space will be needed under the car in the vicinity of the pinion.

and order the correct length cables to hook it all together. I find it more satisfying and usually much cheaper to search out components from late model donor vehicles and put together a custom arrangement.

Most smaller late model vehicles have a handle mechanism that is floor mounted and includes a length of flexible primary cable. These are often used because they allow you some flexibility with regard to where the handle is mounted and the length of cable can be routed to a convenient location under the car to pick up the main activating cables to the backing plates. A typical example of this arrangement is shown in this chapter but there are countless variations that could be used. Find the one that suits best and try your hand at fitting it yourself. It isn't hard.

Some aftermarket suppliers advertise pinion mounted parking brakes that can be useful if you have difficulty fitting a conventional system for some particular reason. That reason might even be simply

CHAPTER 7
EXHAUSTS

The advent of modern heat resistant coatings and the increasing use of stainless steel in the street rodding world has seen the exhaust now being treated as another part of a street rod project where detailing can make an enormous difference.

Custom made exhaust headers are the most common answer to the problem of fitting an exhaust system into the close confines of a street rod engine bay.

Exhausts in street rods have developed from a necessary evil to a thing of beauty in recent years. Once it was sufficient to just mount some pipes to get rid of the exhaust gases and replace them whenever they burnt or rusted out. The advent of modern heat resistant coatings and the increasing use of stainless steel in the street rodding world has seen the exhaust now being treated as another part of a street rod project where detailing can make an enormous difference.

Custom built headers are often required on a street rod because we hot rodders are bent on fitting bigger engines into cars where they weren't originally designed to go. That means space is often at a premium and the only way to get the exhaust away from the engine is to build a snaking, tightly tucked set of tubing headers. It's true that some manufacturers have been able to develop ready made compact headers to suit some of the more common running gear combinations, particularly when fitted into early Fords or Chevys, but step outside the small block Chevy V8 into early Ford or Chevy categories and it is very likely you will have to custom fabricate your own exhaust system.

Sometimes it is possible to retain original factory type cast iron manifolds, such as those commonly used on small block Ford and Chevy engines, but if the outlet fouls on anything in the engine bay it is a major drama to successfully alter one of these types of manifolds. Usually it is simply easier to buy or make tubing headers.

If you can't locate ready made custom exhaust headers to suit your choice of engine into whatever early street rod you're building you will find that

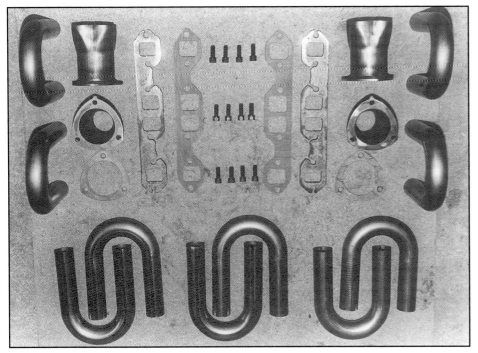

If you can't locate ready made custom exhaust headers to suit your choice of engine into whatever early street rod you're building you will find that most specialist exhaust shops can supply header flanges and bend kits to make it easier to fabricate you own custom exhaust.

most specialist exhaust shops can supply header flanges and bend kits to make it easier to fabricate you own custom exhaust. Some basic welding skills are required but mostly it is just a matter of care and patience when making such exhaust systems.

One thing to watch out for is to make sure there is enough clearance between your exhaust system and any driveline components or steering shafts. Allow at least one half inch clearance around such obstacles, more if you can manage it. You should also do your utmost to keep exhaust heat away from things like brake and fuel lines. Boiling brake fluid or vaporising fuel could cause breakdowns or possibly even an accident in the future.

For the home builder with average tools a neat set of headers can be built using header flanges and a supply of exhaust bends from the local exhaust supplier as shown at left. This set has been painted with heat resistant paint but you can easily have them aluminum sprayed or metallic/ceramic coated for neat appearance and enhanced life.

Abbreviated headers like these will save space under the car because the collector uses space available in the engine bay. This type of header isn't as efficient as a longer more conventional design but it will still flow more efficiently than a stock cast iron manifold. Metallic/ceramic coating such as that provided by Jet Hot or HPC makes them more attractive and easier to keep clean. The engine is a small block Ford V8.

For a daily driver type of street rod conventional slip joints and U clamps may be sufficient to hold the various components together in your exhaust system. However when the rest of your street rod is well detailed a very basic exhaust system can really let the appearance of your car down. It may be worth spending some time to fit some of the better looking flange kits into your system right from the start. It will look much better and it will be easier to maintain in the future.

Of course if your exhaust system is to have much of a future it will need to be made from durable materials or be coated to resist corrosion. Any car that sees irregular use will burn out exhaust systems more readily than one which is used every day so take this into account when making your exhaust system for your rod.

Mild steel exhaust tubing coated with heat resistant paint would appear to be the minimum you should use for your exhaust but for greater longevity

Stock exhaust manifolds are often too bulky or have their outlets in the wrong position for use in a street rod. That's why headers are so often used as a viable option. Now companies like Sanderson headers are offering new manifolds like these that are made to fit the confines of an early engine bay, look attractive and be quiet in operation. This manifold has been further enhanced in the appearance department by the latest metallic/ceramic heat resistant coating, available from companies like Jet Hot and HPC. Engine is a big block Chevy.

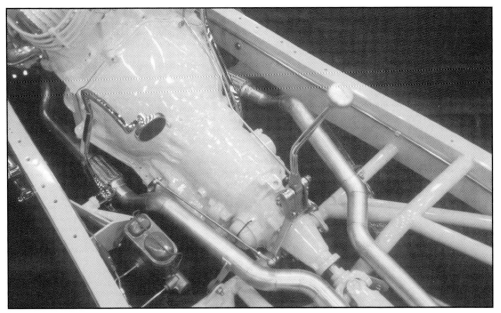

Professionally built stainless steel exhaust systems like this one look fantastic but don't come cheaply. Note the forward mounting bracket near the brake master cylinder. Normally a mount like this with little allowance for movement would break quickly due to the movement of the engine on its mounts but flexible braided stainless steel sections known as bellows have been added to the pipes on either side of the transmission to overcome this tendency.

Here's a better view of the bellows in the exhaust system shown above. You can also get a clear look at the forward exhaust hanger which uses readily available but small and tidy neoprene insulators.

The rear end of the same exhaust system shows how the neoprene insulators have been incorporated again, this time horizontally with welded on hangers.

consider coating mild steel exhausts with one of the modern metallic/ceramic based coatings or having them aluminum sprayed. They not only look good they will lengthen the life of the components considerably.

Avoid chrome plating exhaust system components if you can as they usually discolor and rust too quickly.

For ultimate exhaust detail the modern approach is almost exclusively to use stainless steel tubing and components. It is expensive but you will only have to do it once. However specialist welding skills will be required.

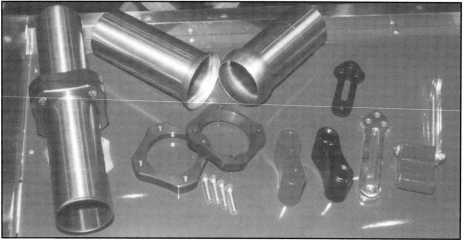

Bassani Exhausts make a selection of profile milled aluminum exhaust hangers with neoprene insulators. They also make a very smart flanged joint that can be welded into your stainless steel system to make it easy to remove in sections for later maintenance.

As you move away from the engine with your exhaust system you will need to incorporate some method of hanger to keep the pipes in position. There's a huge range available from almost any exhaust shop but some of the specialist hot rod after market shops have refined this part of the exhaust system to such an extent that it is now easy to have efficient exhaust hangers that are decorative as well. Where possible though it is a good idea to incorporate hanging brackets that pick up on the flanges of your system rather than having to weld them directly to the pipes. If welded to the pipes you will often find that tension in the system or vibration during operation causes them to crack and break.

One final word on exhausts before we leave you to scan how others have made their custom systems. If you simply run two pipes straight from your manifolds or headers back to the mufflers and then straight on out the rear of the car you may find that the system is difficult to keep quiet enough and it will possibly have a nasty "rap" in the tone. To help overcame this situation you should add a balance pipe between the two exhaust pipes forward of the mufflers. Usually this is added just after the header collectors.

Simple exhaust hangers like the double rubber donut shown here can often be found on late model vehicles. Pick them up from your local auto recycler and the price will be right too.

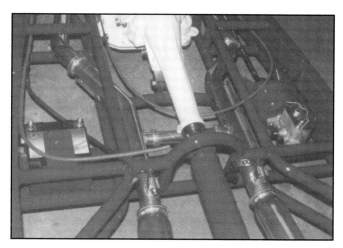

A balance tube linking the pipes together ahead of the mufflers will improve exhaust efficiency and make the sound more mellow. If your exhaust "raps" when you back off the accelerator try fitting one of these connecting pipes.

Now here's a clever way to incorporate an exhaust balance tube, make it part of the existing system. The engine pipes have been drawn together so that they "kiss" each other under the transmission extension housing. This should have the same effect as fitting a balance tube but it looks much tidier and eliminates extra joints where leaks could occur.

It's hard to fault this perfect pair of polished stainless steel exhaust pipes nicely reflected in the equally impressive polished stainless steel fuel tank. Running your exhausts all the way to the rear like this will help minimise "drumming" in the car. Think twice about using turned down ends on your pipes or they will be prone to throwing dust all over your street rod when you start and drive it.

Legal requirements for exhausts will vary from state to state but suffice to say that you should run your exhausts all the way to the rear of your car if possible. At least make sure they are further rearward than the last opening window or you may find exhaust fumes are sucked into the car. I would also recommend that you avoid turned down ends on the tips or you will find the exhaust blows dust and dirt all over your car as soon as you start it.

CHAPTER 8
COOLING SYSTEMS

Cooling systems can be as individual as the street rods they are fitted to.

Cooling systems can be as individual as the street rods they are fitted to. Depending on the particular style of your car you may choose to fit one of the many reproduction radiators that are available or you may elect to have one custom built. A third alternative is to use a readily available radiator from a modern car that can be made to fit your street rod with minimum fuss. Sometimes this alternative can be a much cheaper option.

Reproduction radiators are usually made to fit a variety of common options when it comes to the positioning of the outlets. Usually these types of radiators will be an external duplicate of the original complete with mounting brackets that allow all of your original sheet metal and fittings to be bolted in place. While these radiators may be more expensive than you think you can afford they do have the advantage of being easily fitted for the above reasons.

By their very design most street rods are limited in the amount of radiator surface that is exposed to the incoming air. That surface may have been sufficient for an early, low horsepower engine but it may be a little on the restrictive side to cope with a large fire breathing V8 powerplant.

Should you have a sound original radiator that only needs recoring you may prefer to have a local radiator shop do this for you. However be forewarned that this is sometimes more expensive than buying a repro radiator due to the high cost of some replacement cores and the amount of labor involved in replacing it.

Even if you decide to go with the reproduction radiator option you may find that there isn't one made for your particular car, especially if it isn't one of the more popular early Fords or Chevys. Here's where it can be advantageous to seek out a commonly available late model radiator that can be made to fit in your engine bay, even if it means you

have to make some sort of mounting frame to accept it. Often this mounting frame also needs to be made so that it holds the front sheet metal in place as many early cars relied on the radiator to perform this function.

By their very design most street rods are limited in the amount of radiator surface that is exposed to the incoming air. That surface may have been sufficient for an early, low horsepower engine but it may be a little on the restrictive side to cope with a large fire breathing V8 powerplant. To cope with such a situation you will need to maximise the flow of air through the radiator. This means an efficient fan will be needed to pull the air through the radiator when the vehicle is travelling at low speeds. Furthermore a shroud will quite likely be added to the list of requirements to enable the fan to work at maximum efficiency.

For most efficient operation the fan should be positioned so that half of the pitch of the fan blades extends into the rear opening of the shroud itself. Like water, air will follow the path of least resistance so you need to maximise the radiator's ability to

Above: *Reproduction radiators that are accurate to the originals are even available in aluminum these days. Brackets are included to allow the radiators and all their original type fittings such as the grille shell to fit right up. These premium quality radiators from Speedway are expensive but do make life easy.*

Below: *Here's a trio of reproduction and replacement radiators to suit Model A Fords as made by specialist radiator shop Aussie Desert Cooler. The radiator at far left is a reproduction unit that closely replicates the original item. The other two are to suit Model A grille shells but with alternative single or double outlets to suit different engine applications. These radiators use commonly available cores that have been modified to fit the Model A grille shell. While they don't accurately duplicate the original radiator, once the grille shell is fitted and the hood closed you can't tell the difference. There is a big difference in cost, however.*

Left: *Cooling capacity of the same size radiators can differ dramatically. These two cores are the same size in physical dimensions but the core at the bottom has much more cooling capacity thanks to increased surface area exposed to the air via the larger number and closer spacing of the tubes. The bottom core would be much more suitable for a street rod where radiator exposure at the front of the car is restricted by the size of the grille opening.*

Below: *Here's an example of a late model radiator mounted in an early grille shell. This type of radiator will often be less expensive than a custom made unit but some sort of mounting frame will be required as will grille shell attaching brackets.*

Above: *A late model radiator has been fitted to this '32 grille shell which has been extended to suit. The filler panel at the top will help to force the air through the radiator core rather than let it escape over the top of it.*

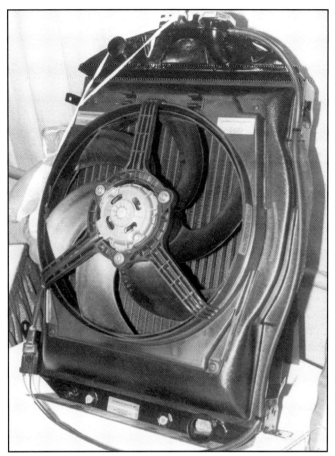

Above: *Clever mounting bracket design has been used to hold a late model radiator, the slim design fan and the headlights on this Model A Ford roadster fitted with a '32 grille shell. Painting the radiator the same colour as the rest of the body helps it to blend in.*

Above right: *The design of this combined shroud and electric fan mount ensures that all the air flow is drawn through the radiator core. The fan is also set far enough into the shroud that it won't recirculate air from outside the shroud. (See text for details).*

Right: *By reversing the polarity of the motor most electric fans can be used either as "pullers" or "pushers". This dual "pusher" arrangement is mounted to the front of a custom made radiator in a Ford Anglia.*

have the incoming air actually pass through it, not around it. If the fan doesn't extend partly into the shroud it can simply draw air from around the outside of the shroud and recirculate it rather than drawing the air through the core of the radiator.

Whereas most late model vehicles fitted with an automatic transmission rely on a cooling system added into the lower tank of the radiator for cooling

Slim aftermarket fans are readily available and they can often save the day when space is a problem. Mount them as high on the radiator core as possible as shown at right for best cooling efficiency. A Thermostatically controlled switche that can be pre-set at your choice of coolant temperature makes for an efficient installation that only works when it is required.

the transmission fluid I would suggest you fit a separate transmission cooler in your street rod. This will help to maximise the efficiency of your radiator because it won't have to cope with the extra heat introduced by the transmission cooler. A separate, dedicated transmission cooler will also do a better job of looking after the operating temperatures of the transmission fluid. There are many options available in this field and most are reliable. Just make sure you mount it where the air can flow around and through the transmission cooler to maximum effect.

If space permits it pays to use an original factory style engine driven fan. This one is fitted to a small block Chevy engine in a '34 Chevy sedan. The fan shroud will enhance its cooling ability by forcing the air to be drawn through the radiator core.

Rather than rely solely on a transmission oil cooler located in the bottom of the radiator it is better to fit a stand alone transmission cooler similar to this example. Mount it in a position where it will have uninterrupted airflow around it and where it is unlikely to be damaged by stones or clogged with dirt.

A wide range of aftermarket transmission coolers are available so it shouldn't be too hard to find one that suits your requirements. The type shown here is among the simplest in design and to mount, requiring only two mounting points and simple brackets. Rubber hoses can be used to connect it to the transmission but make sure you use the correct type.

CHAPTER 9
FUEL SYSTEMS

The fuel system for an average street rod generally doesn't need to be complicated.

Using an original tank is obviously the easiest way to go but before you fit it make sure it is in good condition and doesn't have any leaks. This example is a stock tank in a '39 Ford sedan.

The fuel system for an average street rod generally doesn't need to be complicated. For most street rods you simply need a tank, usually mounted somewhere in the rear of the vehicle and a fuel line to connect it to the engine. However the advent of more modern running gear in street rods does require a slightly more complicated approach.

If you are using a conventional V8 powerplant with aftermarket carbureted induction your fuel system will essentially consist of just a few items. A tank to hold the fuel, a line to supply the fuel to the engine, a pump to facilitate that supply and a filter to ensure the fuel is clean when it arrives at your carburetor.

Using an original tank is obviously the easy way to go in this department but be sure it is in good condition before you fit it. Often such a tank will have scaley rust on the inside, especially if it hasn't been used for some years. You don't want that flaking off and constantly blocking your fuel lines so a thorough inspection and clean is going to be in order. You may even find it is worth your while to coat the inside of the tank with a sealant.

Should an original tank not be available or no longer suitable for your application you will have to pursue other options. This could be as simple as purchasing a reproduction tank from one of the many aftermarket suppliers and these days that tank may not necessarily be steel. Modern plastics have opened up many possibilities even in the street rod world and none more so than in the area of fuel tanks. Here it is simply a matter of purchasing the tank that suits your needs and following the manufacturers instructions.

A second option is to adapt a fuel tank from some other modern vehicle to suit your early street rod. Again there are hundreds of options depending on your particular rod. If it is a roadster or coupe you will find that it is often easy to mount a fuel tank from a modern small car behind the seat. Build a cover around it and nobody will ever know the differ-

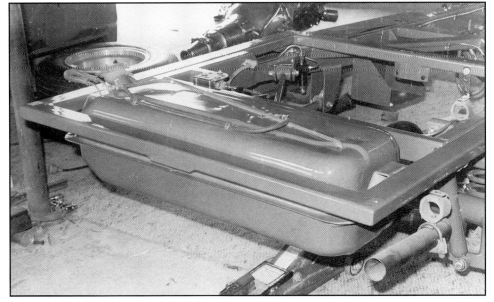

The tank shown here is from a Mitsubishi van but it proved ideal for use in the author's Model A Ford pickup. Stock reinforced mounts at each end of the tank have been used for this installation within the new steel frame designed to hold the pickup bed in place. Even much of the original vent system was able to be retained and the tank already had provision for a return line from the carburetor, which was a feature of the Buick V6 engine used in this car.

A vent tube was fabricated to pick up the original vent outlets from the Mitsubishi tank and hidden within the stakebed hole at the rear of the pickup bed. In some jurisdictions this type of vent that remains open to the atmosphere is no longer legal so you may have to fit a more elaborate system similar to late model vehicles.

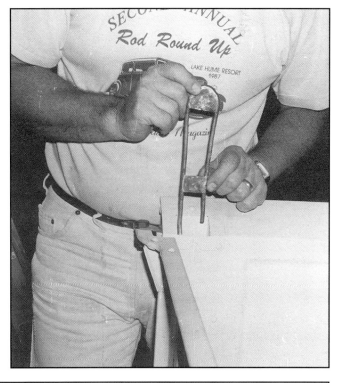

ence. One thing about mounting the fuel tank here is that it is probably also the safest place as it is as far from any point of impact in an accident as it is possible to go. There is less chance of it being damaged in such an accident. However be sure to mount it in the same manner it was in the donor car and don't forget to include some sort of venting system.

For other body styles you will very likely be restricted to fitting any replacement tank in the same position as the original or at least somewhere under the floor. Sometimes a cover can be purchased or made to disguise your replacement tank, especially if

Reproduction fuel tanks are readily available and may not even be steel any more. This '32 Ford example made by "Tanks" is made from high impact resistant plastic but still duplicates the design of the original.

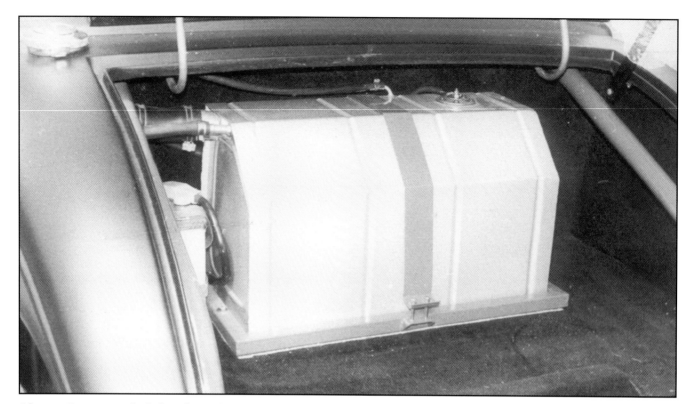

Above: *Custom made fuel tank in a '34 Ford coupe is well secured and vented. Making a custom fuel tank is a specialised task and should only be attempted by an experienced builder. Baffles must be included to reduce fuel sloshing and the tank must be absolutely leak-proof.*

Below: *A wide range of bolt-in aftermarket tanks are now available. Typical examples shown below include a high impact plastic "saddle" tank for Model A Fords and a steel reproduction '32 Ford unit that duplicates the original.*

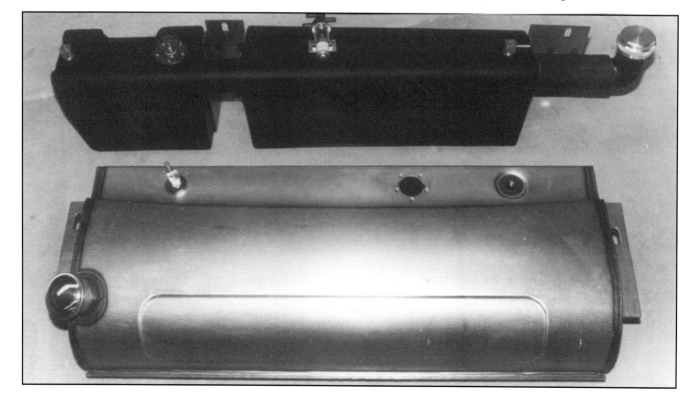

Using a tank from a small late model car is often a good alternative for a coupe or roadster where it can be mounted in this position behind the seat. Make sure to mount it in the same manner as it was originally and route the filler so that any fuel spill when filling can't enter the passenger area. Note the in-line filter used in this installation and the rubber grommet protecting the fuel line where it passes through the floor. Often a tank like this can be fitted and still leave room for a rumble seat and extra passengers. The seat and some removable panels will hide it all from view when finished.

the original actually formed part of the bodywork of the car such as on a '32 Ford. Again just be sure the tank is mounted in the same manner as it was originally and be sure to vent it in some manner.

Until the advent of smog reducing regulations most fuel systems relied on a simple venting system that allowed the fuel vapor to escape to the atmosphere without allowing the fuel to spill from the same point. In some parts of the world such a system is still acceptable on older automobiles but you will find in many instances that fuel vapor is no longer permitted to be vented to the atmosphere. It must first be routed through the engine itself via a charcoal canister or other entrapment facility. While it may seem like too much hassle to incorporate such a system I like to look on these things as just another challenge that we hot rodders like to overcome.

Modern fuel systems have a few other "extras" that older hot rodders haven't had to deal with before. Fuel injection systems and some of the later carburetor installations require a return line that takes excess fuel, over and above the engine's immediate needs, back to the fuel tank. This is no big deal, it just requires a duplicate fuel line for the purpose.

Fuel tank filler necks can be the cause of frustration come filling time if you don't include some form of ventilation system that runs alongside the filler.

Here's another example where a late model fuel tank has been adapted for use in a Model A Ford roadster. These tanks are often quite narrow, consequently using up minimal space, but they still have ample capacity for a roadster. Access to the fuel sender and pickup is very good and the tank could be removed quite quickly if necessary in the future.

Basically all that is needed is a second tube that allows air to escape from the top of the tank to the top of the filler neck so that when the fuel is rushing in it doesn't cause an air lock and consequently spurt back at you. Study modern fuel tank fillers for neat ways to incorporate such a vent.

When it comes to fuel pumps the process is again quite straightforward, You have the choice of a manual pump as fitted to the engine you are using or if that isn't possible there are many aftermarket electric pumps available. Sometimes the original mechanical pump has to be deleted from an engine to provide clearance for other mechanical components such as steering box or chassis rail. It's no big deal, just use an electric pump.

Mounting aftermarket electric pumps is usually very easy as they will generally include the mounting bracket and screws, etc. Put a little thought into where you mount it though. Make sure it is accessible after the car is finished and that it won't foul on any other fittings. Keep it well away from exhaust fittings and where it won't allow fuel to spill onto exhausts if the system develops a leak at some point in the future.

It is generally accepted that the best place to mount an aftermarket electric fuel pump is at the rear of the vehicle in close proximity to the tank.

Many modern fuel injection systems require a pump that is housed inside the fuel tank or at least a

Typical aftermarket electric fuel pumps come with their own mounting hardware which makes installation easy. Mount your fuel pump where it is easily accessible and away from exhaust components.

pump that is capable of operating at very high pressures. Make sure you are aware of any such requirements before you get too far into your project and consult a fuel injection specialist if necessary.

Fuel lines should be treated in much the same manner as brake lines. That is, they should be securely mounted every 12 inches or so and routed away from contact with other components. Be sure to

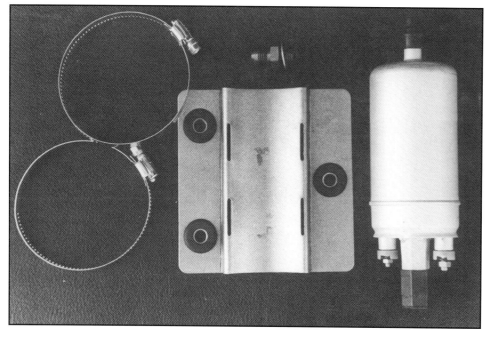

Modern fuel injection systems often require a fuel pump that is immersed in the tank or at the very least a pump that is capable of delivering at very high pressures. An aftermarket example of the latter is shown here.

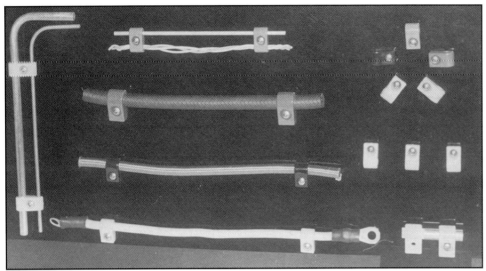

Left & below: *As with brake lines, your fuel lines should be secured every 12 inches or so along the chassis rails and cross-members. Many different styles of retaining clips are available and some are dual purpose as shown here in that they can be used to secure brake and fuel lines or brake lines and wiring. Keep the fuel lines to the inside of the chassis rails and cross-members where possible as shown below.*

avoid running them along the bottom of chassis rails where they could be damaged by flying stones or where they could come into contact with floor jacks during maintenance. Keep them away from the heat of the exhaust system or you could find a vapor lock problem emerges later on.

When it comes to filters there are countless varieties available for you to select from. Just be sure the one you select has the capacity to flow enough fuel for your application and mount it where it is easily accessible.

Here's a clever fuel filler that has been built into a Model A Ford Coupe body. Note the neat little button release for the rumble seat hidden inside the filler recess and the drain tube to prevent any spillage escaping from the recess area or entering the passenger compartment.

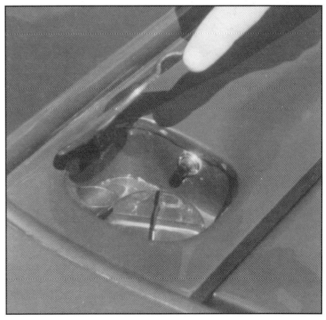

Dual fillers are a feature of this tidy tank installation in a Model A Ford roadster pickup. A compartment built into the forward area of the pickup bed looks neat but the access door still makes the actual tank and possibly the battery easily accessible.

Unless your tank has a short straight filler neck like this one you will probably need to include some form of bypass to allow the air out of the tank quickly as the fuel goes in. Otherwise the tank will be prone to splashing back when filling with annoying consequences. Normally such an air bypass would come from the top of the tank near the filler neck and join back into the neck near the cap. (See top photo on page 108).

CHAPTER 10
WHEELS AND TIRES

There are some basic safety requirements that you should heed when it comes to selecting wheels and tires no matter what car they are going on.

Wheels and tires are subject to such individual tastes and are so dependent on the style of your street rod that there is little I can do to influence your particular choice. However there are some basic safety requirements that you should heed when it comes to selecting wheels and tires no matter what car they are going on.

It is my opinion that no car with the performance potential of the average street rod should be fitted with wheels that don't have safety beads to retain the tires. That ordinarily wouldn't be a problem since all modern wheels incorporate this feature. However the desire to build authentic nostalgia style street rods sometimes incorporates the use of early wheels that may not have safety beads. Don't compromise your own, or the safety of your passengers by using such wheels. It is possible to purchase or modify wheels so that the nostalgia style is retained but the safety beads are incorporated.

Another way that safety is often compromised in this area is to use wheels with too much offset. This can put extreme loads on bearings and axles that leads to premature failure, sometimes with disastrous consequences. This is the very reason many vehicle

There are so many different types and styles of wheels available now that it can be quite confusing just making up your mind which type you will use on your street rod. Selecting a style of wheel is sometimes better left until most of the other styling elements of your car are in place, then go with the wheels that suit best.

Avoid using old wheels that don't have safety beads. These ridges keep the tire seated on the rim and resist the tendency for the tire to roll off its seat in hard cornering which can cause sudden deflation. All modern wheels have this feature but many cars built prior to World War II had rims that did not have a safety bead.

If you want to retain that early wheel appearance several manufacturers now offer brand new wheels that accept all the original style hub caps and trim rims. These types of wheel accessories are also widely available in reproduction form.

When you are not sure whether a particular wheel is going to be suitable for your street rod it is often a good idea to get hold of a similar wheel without a tire fitted and bolt it on place. This will allow you to check suspension, brake and fender clearance.

codes have limitations on widening the track width on vehicles. It isn't done just to annoy auto enthusiasts, there are safety implications involved. If you aren't sure what the limitations are you will usually be fairly safe if you only use wider wheels where the increased width is equally inward and outward when compared to a standard wheel off the axle you are using. Certainly you should be wary of using a wheel that widens the track of your rear axle by more than one inch each side and the front track by half an inch each side. Track width is normally measured from center of tread to center of tread.

Tradition demands that street rods usually have wider wheels on the rear than on the front. Again there are safety reasons for not going too extreme in this area. Skinny fronts and wide, wide rears may look tough but in other than perfect conditions this combination can result in bad handling characteristics and poor braking. To minimise potential danger but still keep with that traditional "bigs and littles" look I would suggest you keep the front wheel width

Severely offset wheels like these should not be fitted for normal street use as wheel bearing failure or axle flange breakage can occur. A wheel with this much offset requires expensive full floating hubs.

To accurately measure your projected wheel spacings start by measuring from the bolt flange of the rear end to the outer edge of the fender. A straight edge across the fender opening provides a reference point to make measuring easy.

Next place the straight edge across the axle flange and measure the distance to the inner edge of the fender well. These measurements will tell you how much back space and front space you have available and how wide the wheel can be to fit inside the wheel well. Remember to allow for the bulge of the tire side wall and allow at least one inch clearance from the tire side wall to the inner edge of the wheel well for suspension movement.

to a minimum of 60% of the rear wheel width. Tire sizes can also be juggled a little in the same manner to give you a combination that still looks okay but doesn't compromise safety at the same time.

Modern tires are usually radial ply construction and they are a vast improvement over the old bias ply construction that was popular up until the late sixties. Avoid mixing bias ply and radial tires

Be observant for other fittings in the wheel well area that could have an affect on your wheel size. It would be easy to overlook the fact that the fender support brace here will be very close to the tire if extra clearance isn't provided.

together on your street rod and I would strongly recommend you only use radial tires on a regularly driven street rod. Bias ply tires may be "right" for your nostalgia rod but performance and handling will be far inferior to a good set of radials.

There is always a temptation to apply the "more is better" equation to rear tires on a street rod. While it may fulfil your desire to have a mean looking hot rod you may also be adversely affecting the handling and ride of your car at the same time. One of the major negative affects of ride quality is unsprung weight at the rear end. Our preoccupation with brute strength sees a vast number of street rods fitted with nine inch Ford rear ends that add more than enough unsprung weight to this area without adding to it unnecessarily with the biggest tires we can find. Compromise enough to find a tire that fills the wheel

Measuring the front wheel spacing can be a little more difficult because the fender doesn't extend as far around the wheel. Using a spirit level as a straight edge allows you to gain an accurate front space measurement. Check carefully for possible interference with suspension and steering components and don't go with excessive wheel offset on the front or you risk increasing the likelihood of bearing and/or spindle failure. Steering can also be adversely affected by excess offset. Be guided by the original factory offsets and wheel manufacturer's recommendations. Also check caliper clearance, especially if using smaller than 15 inch diameter wheels.

The result of all that measuring before you order your wheels is a wheel and tire combination like this. The wheel and tire fits the wheel well without interference and fills the space without looking too big. Keeping an even gap all the way around between tire and fender is the secret to making a street rod look balanced.

You may think huge tires like those below look really cool but before you opt for a combination like this be sure you are aware of its limitations and legal ramifications. Ride and handling will be compromised and a wheel/tire combination like this may not be legal in your state.

well sufficiently rather than one that stuffs it to the extreme and makes your rear wheels look like thimbles. You will be happier in the long run.

The pictures accompanying this chapter show how to go about measuring the front and back spacing of wheels for your street rod. Most wheel manufacturers can supply wheels with the center positioned in the right place to suit your vehicle. Take the measurements as shown here and then visit or call the wheel supplier you intend to use. If they can't supply a standard size wheel that comes close to your requirements they will very likely be able to build custom wheels to suit. Make sure you follow the manufacturers recommendations when it comes to wheel nuts and don't compromise on quality in this critical part of your street rod.

CHAPTER 11 ACCESSORIES AND SAFETY ITEMS

Take the time to include as many safety related items in your street rod as you can. If you only ever use them once they will have been worth the effort.

Accessories and safety items have an important role to play in your street rod. It has taken a long time for hot rodders to overcome the reckless daredevil stigma that surrounded the hobby in its early days. For the most part it is now generally accepted that rodders are very safety conscious but sometimes in the rush to finish a project some important things are overlooked or skimped on. Take the time to include as many safety related items in your street rod as you can. If you only ever use them once they will have been worth the effort.

The types of things I am referring to are now almost standard fare in modern cars but there is no reason not to include them in your street rod. Things like four way flashers, retractable seat belts, burst proof door latches, two speed wipers and battery isolator switches are just a few of these items.

Back when the cars we use to build our street rods were originally made by the factory they never included much in the way of safety equipment. Consequently there is no provision for such items as seat belts. The mounting of seat belts is not something that should be taken lightly. You may or may not agree with having to wear them but in our part of the world the evidence in favor of them is overwhelming with a huge reduction in road deaths as a result of their compulsory application since the mid 1970s. Make the effort to include them but be sure they are mounted properly.

To do so will require some re-engineering of things like door pillars and the incorporation of floor or chassis mounts. Some early door pillars weren't much more than a light steel channel with a wooden insert. This is not sufficient to support an upper seat belt mount. Reinforce the pillar with a steel insert or better yet turn the entire pillar into a boxed steel tube for its entire length. Where the upper seat belt mount is located the mounting plate should be substantial and well reinforced. Don't compromise in this area.

At the floor level seat belt mounts should pass through a steel crossmember or reinforced steel body frame member. Don't just bolt them to a wooden floor member or thin floor sheet metal. At the very least a large flat reinforcing plate should be included where the mount is through a body frame member or a sheet metal floor. If involved in an accident the forces exerted on these mounts can be huge, you don't want them to pull out.

Placement of the mounts in relation to the seat is also important and there is a diagram included here

The author even managed to fit retractable seat belts in a Model A Ford pickup where space is really at a premium. Note how the door pillar has been extensively reinforced with steel in the upper seat belt mounting area. Lower mount goes through the rear body frame member.

Left:
The addition of seat belts in your street rod can be achieved in a tidy manner as shown in this '34 Ford tudor. Study the way it is done in new cars and pick the style that suits your street rod best.

Above:
A shelf mounted retractable belt has been fitted into this street rod by moving the seat forward and incorporating a new shelf where there wasn't one originally. This makes for a very tidy installation but the belt is still easy to use. Radio speakers take advantage of the new shelf as well.

Left:
The rear seat in this model A ford has been moved forward to provide working space behind it for the retractable seat belt mechanism. Note the strong steel mount built into the body frame and linked into the rear body seam for maximum available strength.

There are scientifically tested mounting areas for the securing of seat belt anchorages. All measurements for seat belt mounts are taken with the seat in its rearmost position and with the seat back at its "design angle", usually 15° from vertical. A complicated formula can then be used to determine where the upper mounts should go.

Point R in the diagrams represents the position where the torso joins the leg and is known as the Seating Reference Point (SRP). This is arrived at by taking the Seating Reference Plane (front view through the center of the particular seat) and measuring forward 70mm from the squab and up 95 mm from the base of the seat.

Lap anchorages must be at least 165mm apart and in side view must be between 25° and 80° to the horizontal when measured from the pelvic reference point. It is impossible to precisely determine this point on a human so a tolerance of + or - 10mm is typically made. However the scale drawings of the anchorage locations should allow for this tolerance, hence it is prudent to keep any belt mounting location 10mm inside the shaded area if possible.

The location of the upper anchorage is determined by a complex three dimensional formula and is dependent on how far Dimension "S" is located sideways from the SRP. The two scale drawings show the allowable area in a side view when the anchorage is located (a) 300mm and (b) 200mm outboard of the SRP.

Unless "S" is less than 200mm, the anchorage must be at least 450mm above point "R". FK is at 120° to RL with point "B" located 260 + "S"mm from point "R". FN is at 65° to RL, with point "C" located 315 + 1.6 x "S"mm from point "R". NJ is a vertical line, with point "J" located 1.3 x "S"mm from point "M".

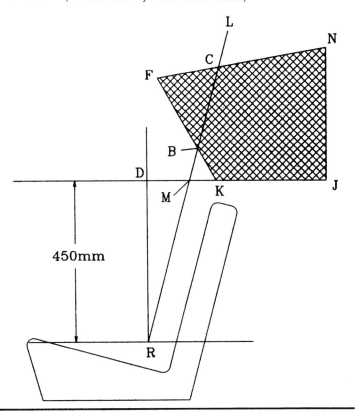

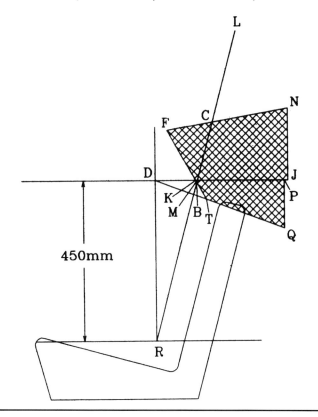

Clever use of the interior trim allows the belt to disappear behind the side panel in this Model A Tudor. The armrest built into each side obviously also provides additional space for the retractable belt mechanism.

to guide you in this respect. Mounting your seat belts in the area defined by the diagram is the safest position for average sized people according to scientific calculations on which new car design rules are based.

A battery isolator switch is one accessory that I would not leave out of a street rod. You have a large investment both in time and money in your street rod and one of these switches will, if nothing else, give you great peace of mind. It can also act as a secondary security fitting as anyone trying to steal your street rod would have to know where the switch is and have a key to it in order to start the car.

Fit the isolator in the positive feed cable of your electrical system but try to keep all your cables as short as possible to minimise voltage drop. Be forewarned however, that some modern engine management systems currently finding their way into street rods require a constant electrical feed to maintain their memory. To facilitate this you may need to have a small bypass electrical feed wire to the computer that isn't shut down by the isolator switch.

Driving your street rod in the rain may not be your idea of fun but if you intend to use it even semi-

A roll bar is often a worthwhile safety addition to a street rod, especially when it can be incorporated as neatly as this one in a '32 Ford Coupe. It also provides a secure upper mount for the lap-sash seat belts.

Space is always at a premium in an early car so make the most of it where you can. Here the area behind the seat has an accessible compartment where the spare wheel, windshield washer bottle and battery are located. A clip in panel hides it all from view.

Battery isolator switches are a good idea in any street rod both from a safety and a security aspect. This one has been fitted in the seat riser of the author's Model A pickup where it is out of sight once the seat is in place but it is still easy to reach and operate.

Modern wiper systems can be fitted to street rods without being obtrusive. This triple wiper system is from a Nissan Patrol four wheel drive vehicle but was easily adapted to the Model A. Because the blades park parallel to the top of the windshield and the arms are hidden under the sunshade the entire installation isn't seen from a normal viewing angle. In operation this is a very efficient, two speed system that cleans almost the entire glass area with each wipe.

Here's that same wiper system viewed from inside the car during installation. The whole arm drive mechanism fits inside the Model A header panel and was easy to install. Each blade drive has its own pedestal mount that bolts in place through the header panel. Drive arms have been shortened to suit.

regularly you are going to need a decent set of wipers. This is one aspect of street rod building that is often overlooked. A wiper is needed for registration so any simple accessory wiper is added on at the last minute or the inefficient original system is simply bolted back in place. You will be much happier if you take the time to fit a modern efficient system and engineer it into your car. Granted some wiper systems can look ungainly, particularly on a roadster or tourer but even in these situations it can be overcome with a little forethought. There are examples of adapting late model wiper systems to early closed cars shown here but these are examples only. Use your own ability to find a system that you can make

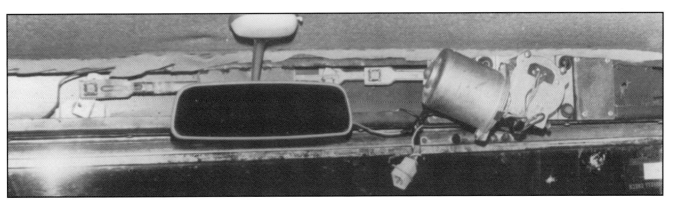

All that was required to fit the drive motor for the Nissan wiper system was to make small adaptor plates and pop rivet them to the header panel. The motor mounting plate then screws to these adaptor plates. The drive arm sitting here at an angle fits directly to the motor drive.

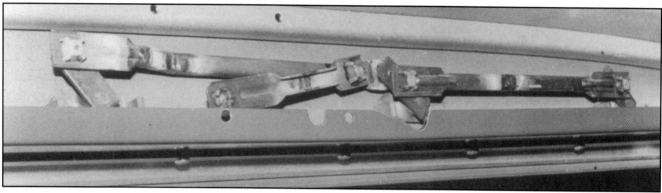

Here's the same Nissan system fitted into the header panel of a '30 Model A closed cab pickup. This part of the header panel is normally solid wood but this one has been rebuilt from steel in a similar manner to early Model As. This provided space for the wiper system. Note how the arms simply press onto their drive pivots using nylon button inserts. For this installation the motor was mounted in the center of the header panel as shown at left. Mounting it here means the driver or passenger's head is less likely to come into contact with it in the event of an accident. The lower left photo shows how the motor has been mounted as far forward as possible to minimise the intrusion of a covering header panel. Wiring is passed down the inside of the windshield pillar.

work in your own car. When it comes to open cars the ungainly appearance of a wiper perched on top of an otherwise clean windshield frame can be overcome by incorporating the wiper mechanism into a removable roof. You won't be driving in the rain without your roof so apart from when the roof is on the wipers shouldn't be needed.

Study wiper blades on other vehicles too and visualise ways they could work on your street rod. No longer is it necessary to have a blade resting across your windshield at an angle that could detract from the clean appearance of the car. The examples

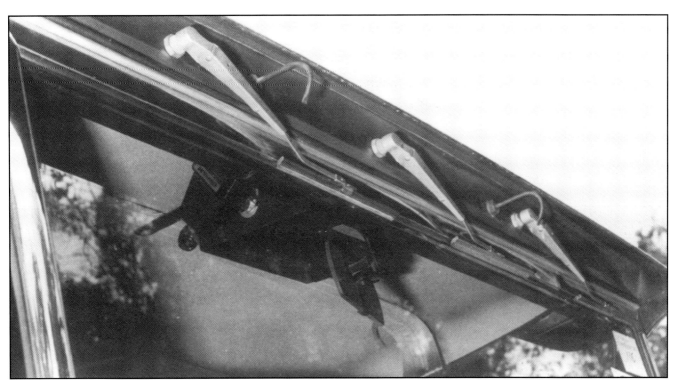

Here's another variation on a late model triple wiper system fitted to a '32 Ford closed cab pickup and using components from a BMC/Leyland vehicle. Note how washers have been incorporated into the header panel as well on this one. The drive system utilizes a worm drive cable housed in a tube that can be routed around gentle curves as shown below, allowing the motor to be mounted out of harm's way in the rear corner of the cabin. Many BMC/Leyland products used this type of wiper system through the '50s and '60s and many of the components can be mixed and matched between models to make your own hybrid system.

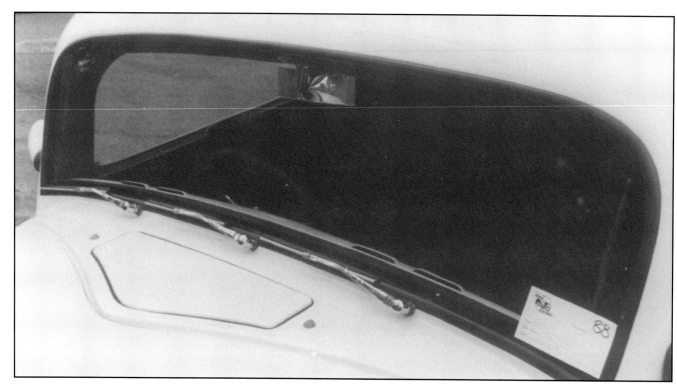

Late model triple wiper systems can even be made to look good when they have to be exposed as on this '34 Ford tudor. The windshield has been converted to the modern "glue-in" style and wiper blades selected that park neatly across the bottom of the glass. Note the unobtrusive washer outlets mounted either side of the cowl vent and the neat louvers added to the upper dash rail to act as demister outlets. You can add all these safety accessories without spoiling the antique character of your early street rod if you plan carefully and hunt for the most suitable components.

Here's a neat wiper arrangement built into a removable hardtop for a roadster. Since the wipers come off with the roof the top of the windshield is left clean and uncluttered. All that is needed to allow this to work easily is to have some form of quick disconnect fitting for the electrical wiring.

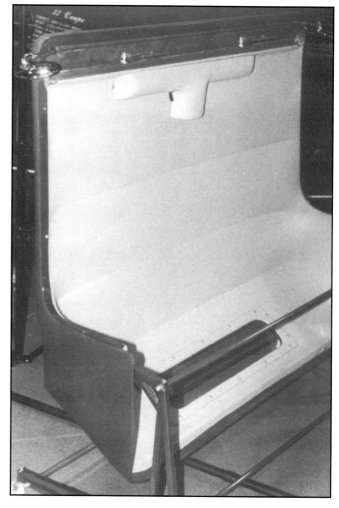

shown here use blades with an angle on them that allows them to park at the top of the windshield and parallel to it. Consequently on the Model A Ford they are fitted to, the wipers are barely even seen when they are in the parked position. If you include the wiring harness and switching from these later wiper systems you will often find that they include the two speed mechanism and the self-parking ability that makes them such a smart installation. The first time you have occasion to use the wipers you will appreciate the effort it took to fit them.

Even where space is at a premium as in a Model A Ford you can still fit a demister if you try. This system uses custom made outlets that feed into the area at the bottom of the windshield that used to be part of the original ventilation system. Note that the center openings have been blanked off effectively turning the dash rail area into a long box that opens to the base of the windshield. Once the dash rail cover and kick panels are in place none of this demister system or its feed tubes is visible from the driving position.

The windshield fame has a broad flange at the lower edge which extends back under the top of the dash rail cover. This flange is drilled in three places to allow the demisting heat to come through to the glass

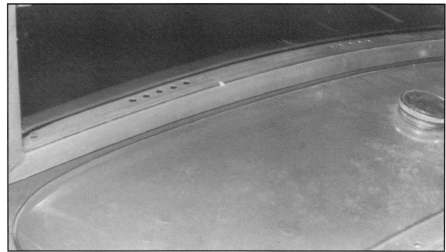

Finally in those instances where the wiper blades can't be made to "disappear" from the viewer's glance, consider removing them when the vehicle is parked at a show or run and fit little coloured covers over the drives. All of a sudden your street rod will look much more attractive and the casual onlooker won't even realise why.

It's nice to enjoy the benefits of modern safety conveniences like demisters and windshield washers and there is no reason why you can't fit them into most early street rods as well. I like to respond to the challenge of finding a way to fit these convenient accessories in a street rod without making it obvious that they are there. Fitting demisters in particular to an early car can be quite a challenge, especially cars like Model A Fords where the space between dash and windshield is very limited. Nevertheless it can be done as I have shown here. Many of these examples are in my own Model A Ford pickup and since finishing the vehicle several years ago I have had occasion to need almost all of them.

Shortly after the pickup was finished I came upon a fellow rodder one night who had lost a wheel off a trailer and the trailer was left sitting in a dangerous position on a bend in the road. I was able to park the pickup further up the road with its four-way emergency flashers turned on to warn other motorists of the danger ahead. This is just one example of a

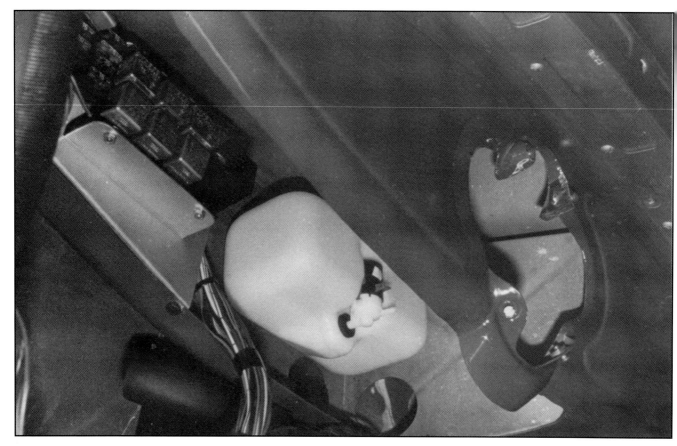

benefit gained from taking the trouble to incorporate modern safety items in your street rod. There are many more and if you are an experienced hot rodder I'm sure you can cite many of your own.

Even in their original form early car door latches weren't the most efficient of designs. The old tongue and striker system is more suited to a stationary building than it is to a moving vibrating automobile. Many readers will have experienced the unnerving feeling of having a door fly open on an early car or at least noticed how much they bounce around in their opening. This can be overcome and your street rod made much safer with the inclusion of burst-proof or bears claw door latches. This is another item that you will be thankful you took the trouble to fit after your street rod is finished and on the road.

Several examples of burst-proof door latches fitted to street rods are shown in this chapter. Many versions are readily available from the street rod aftermarket but equally you will find many that are suitable if you check out almost all modern small cars. Normally the latch is fitted to the door and the post section into the door pillar but it doesn't have to

The bottom of the original Model A fuel tank has been cut away to open up valuable space for mounting important items like the wiring panel and the windshield washer unit. This washer unit is filled through the original Model A fuel cap in the cowl so you wouldn't even know it is there. The outlet is mounted just behind the hood hinge where it is hardly noticed.

be that way. Sometimes it is simpler on early cars to fit them the other way around. It doesn't really matter as long as you can incorporate a release mechanism that is accessible and easy to use.

One other accessory that is also a safety item and one that I believe should be fitted to every car is a fire extinguisher. You will only ever have to use it once to recognise what a good idea it is to have one fitted in your street rod. Attention should be paid to where the fire extinguisher is fitted in your car. Ideally it should be somewhere within reach of the driver when sitting in the drivers seat and with a seat belt on. That way you can reach it even if trapped in the car in an accident. Unfortunately I know of an instance where a life could have been saved if this had been the case a few years ago. The car concerned had a fire extinguisher fitted but the driver couldn't reach it and he was trapped by the legs.

Look at mounting the extinguisher immediately in front of the drivers seat or between the front seats if there is room. Next best mounting position is on the kick panels or firewall/under dash area but even if none of these positions is practical at least mount one somewhere in the car. It might not be your own life or car that you save but someone else's because you had the forethought to include a fire extinguisher in your street rod.

It can be difficult to fit burst-proof door latches into narrow early doors like this Model A Ford roadster. The owner of this one added a feature panel to the interior trim that also allows more room for the mechanism inside the door.

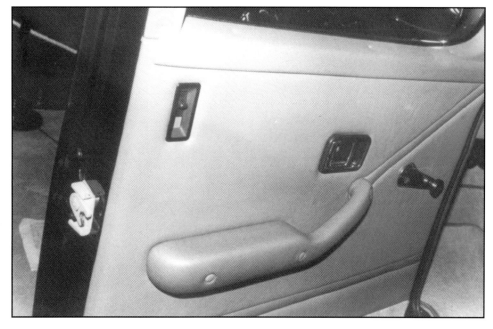

Not only has a late model burst-proof latch been fitted to this '34 Chevy door but also the inside release handle and the modern style slide lock mechanism. The end result is a tidy installation that doesn't look out of place and is functional as well. A stylish arm rest adds another comfort accessory to this nice interior.

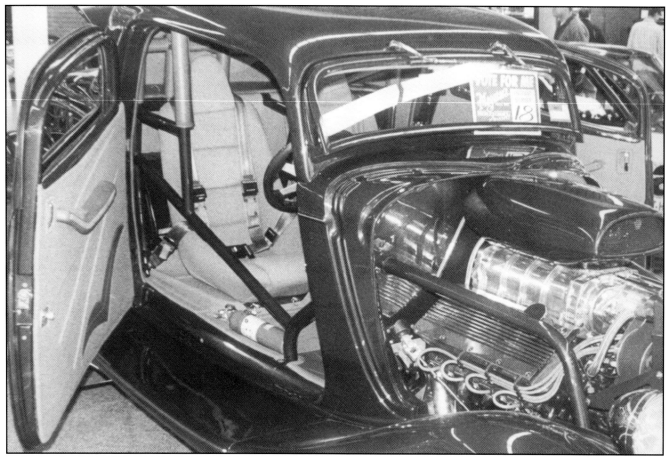

This dual purpose street rod sees both street and strip duty so it incorporates many extra safety features such as the full roll cage with padded upper body area, full harness seat belts and a fire extinguisher by the driver's seat.

Every street rod should have a fire extinguisher fitted somewhere in the passenger area, preferably as close to the driver as possible. This way it can be reached quickly in an emergency

CHAPTER 12
ADVANCED ENGINEERING FOR STREET RODS

Elaborate, advanced engineering techniques for radical street rods are not something the novice street rod builder should attempt without gaining many years of expertise first, or consulting someone who has that experience.

The stepped in portion of this chassis has been well engineered in that it includes large torque boxes at the junctions. Legs of the center X member link into the same point making it all very strong.

The primary purpose of this book isn't to turn the reader into an experienced automotive engineer in one easy lesson. Consequently most of the content has been deliberately kept very basic. Elaborate, advanced engineering techniques for radical street rods are not something the novice street rod builder should attempt without gaining many years of expertise first, or consulting someone who has that experience.

Some aspects of building even a simple street rod do require expertise though and I have tried to include them in this book. The purpose of this chapter is to give the reader a glimpse of some of the more common modifications that require extra engineering expertise currently in vogue in street rodding and to at least point you in the right direction if you want to incorporate one or more of these modifications in your own street rod.

Gaining a low ride height is probably the number one priority for current street rodders but this often requires fairly elaborate changes to the chassis or suspension components. There are a few examples of such modifications shown in this chapter. Study them closely and always err on the side of safety rather than just good looks.

Raising the centre section of front crossmembers for transverse sprung front ends is one way to easily and quickly lower a street rod. However usually this will mean the side rails of the chassis will need to be notched for extra clearance. Don't go overboard and make sure the notched area is reinforced adequately in the process.

At the rear end of most early street rods the amount of kick up in the chassis is often not sufficient to allow the car to ride as low as the owner would like and still retain adequate suspension clearance. Taking a "C" shaped scallop out of the lower half of the chassis is one way to achieve the extra clearance required but again err on the side of caution. Be sure you aren't reducing the strength of the chassis too much. Ideally such a modification should not reduce the depth of the chassis rail by any more than about one third and it must be boxed on the inside to maximise its strength.

Another way to gain more suspension clearance in this area is to completely recontour the chassis rail so that it has more kick up. This modification also involves major floor and lower body reconstruction and shouldn't be attempted by the amateur. However when engineered properly by an experienced operator it does have the advantage of being almost undetectable.

Stepping the rear end of a chassis up under the rear seat area is another way to gain a low ride and maintain suspension clearance. This application is often required on late '20s ladder style chassis such as those used under Model A Fords. Several examples are shown in earlier chapters of this book as well as those shown here. As long as the step up isn't too radical this is a process that is not hard to execute but you must be sure to maintain and enhance the overall strength of the chassis. All welds should be

Raising the centre section of an early Ford front crossmember will lower the vehicle but you may have to notch the chassis side rail as shown here for spring clearance. Ensure that at least two thirds of the original rail remains and reinforce the area to compensate for the piece removed.

In order to gain a low stance at the rear the chassis of this vehicle has a "C" shaped section taken out and reinforced above the axle housing. This compensates for inadequate kick-up in the chassis but care must be taken not to reduce the width of the remaining chassis rail by too much. The next option if even more clearance is required is to rework the shape of the chassis to incorporate more kick-up. However major floor modifications will be required if this is the case.

Another stepped in rear end where the transition points have been very well reinforced to withstand the extra loads it will be asked to endure with the huge rear wheel/tire combination being used. Be wary of using such large wheels and tires for general street use as ride and handling will be compromised. Four link rear end locating system features a diagonal support bar which performs the same function as a panhard bar. Unsprung weight and forces exerted on all suspension components in this type of rear end arrangement will be greatly increased. Workmanship must be of the highest quality.

Step up at the rear of this reproduction Model A Ford chassis is quite severe and will require extensive reworking of the floor. Note how all welded joins have been fish plated for maximum strength. You may consider this over-engineering to some extent but there is no doubt that the joins are unlikely to fail. Triangular gussets added inside the rear corners would add even more strength to this area. The end result is a rear suspension that has plenty of travel so ride quality won't be compromised and the rear end housing won't bottom out against the chassis.

A simple method of narrowing the rear of a chassis is to cut off the rear legs of the chassis, weld a member across the remainder of the chassis and then weld the original legs back onto this new crossmember, but closer together than they were originally. Narrowing a chassis in this manner creates problems as shown in the diagram here and should be avoided. A better solution is shown below in option 1.

Highly stressed section.

Driveshaft clearance required which will weaken structure.

Option 1

Torsion box and substantial rear crossmember similar to late model GM front subframes

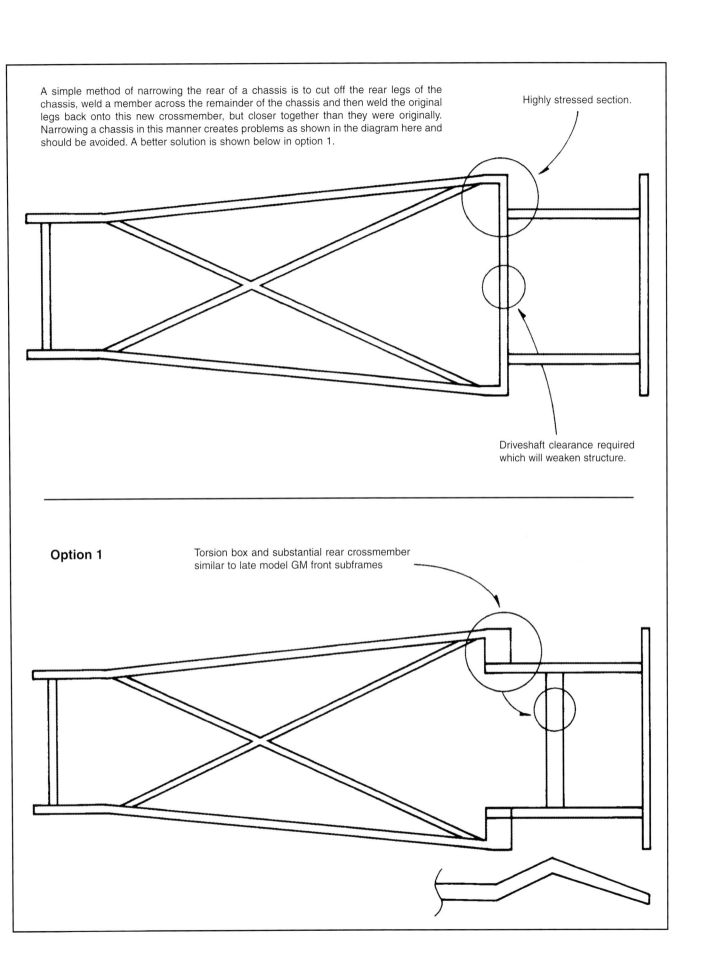

Really huge rear tires are too big for comfortable street use and can be prone to aquaplaning in the wet. Before you build this style of hot rod be sure you understand its limitations.

carried out by a qualified or experienced welder and where possible gussets should be included.

Usually floor modifications will be required when this modification is used but with careful planning and keeping the step up on the conservative side the amount of modification can be minimised. Study the examples in this chapter for some further guidance.

Many early chassis are not only restricted in the amount of suspension movement available but they are also restricted in how much space is available inside the wheel wells for the larger wheel/tire combinations we hot rodders like to use. Widening the fenders is one option to overcome this problem but go too far in this department and the car will look unbalanced. The other option is to step the chassis rails inwards just in front of the rear end to provide a bigger wheel well. Body modifications will be required to the inner fender area as well.

This is one type of advanced chassis modification that is often done badly. If not carefully designed and executed a stepped-in chassis can be left much weaker than a boxed original. There are examples shown here to guide you if this is a modification you need to make to your street rod. The most important area is the junction where the step-in occurs.

Radical independent front end with hidden suspension is a work of art but not something that should be attempted by the first time builder.

Another radical independent front end that will feature hidden suspension when completed. Again this is not something that should be attempted by the first time builder. With this suspension arrangement cantilever rods from the lower outer suspension arms activate coil-over shock absorbers that will be hidden behind the grille. The shock absorbers are minus their springs in the photo. The solid tie rod is only there to hold the wheels in place for display as a steering system is yet to be fitted.

ABOUT THE AUTHOR

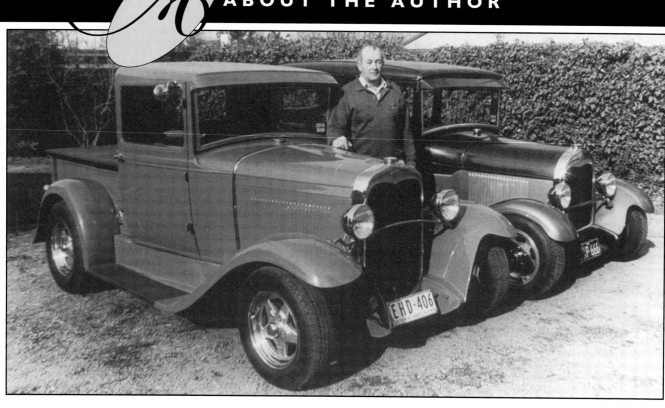

Larry O'Toole is the publisher of Australian Street Rodding Magazine which he established in 1976. He is also a "hands on hot rodder" who is blessed with the ability to do most of the work on his own street rod projects including bodywork and paint. Raised on a wheat farm at Ultima, Victoria, he moved in 1973 to Castlemaine, a hot bed of hot rodding activity where he now lives with wife, Mary and their four children.

Larry currently has two finished street rods including a Model A Tudor that has been on the road for more than 22 years. It is big block Chevy powered and runs on L.P.G. (propane) which is popular as a cheap alternative fuel in southern Australia. His other street rod is a '30 Model A Pickup that is Buick V6 powered and has been in active use since 1990. Both cars are almost entirely home-built in Larry's own workshop, the source of many technical articles for Street Rodding Magazine and for this book. Other ongoing projects include a '32 Ford Hiboy Roadster and a '36 Ford Flatback Tudor.

Previous books on hot rodding compiled and edited by Larry O'Toole and published by Graffiti Publications include the titles; How To Build Your Own Custom Street Car, Street Rodding Gallery, Street Rods In Color, the Colorful World Of Street Rods, Styling Street Rods and Nostalgia Street Rods.

ACKNOWLEDGEMENTS

A book like this doesn't happen on its own. A great deal of the content has come from our Street Rodding Magazine files compiled over the past 22 years. Others have had considerable input into those files in that time but in particular I wish to single out Colin Hall and Des Kelly who were responsible for taking some of the photos you see here. Other contributors to our files included Darryl Poulsen and Tony Parker.

For many years Ted Robinette has acted as the Tehnical Adviser for our magazine and his knowledge and direction is also reflected in this book. I would like to thank Chris Maxwell for his input and advice as the book was nearing completion.

Much of the chapter on brakes that appears in this book is reproduced from an article in Australian Street Rodding Magazine that was originally written by Warren Mills. Credit goes to him for his knowledge in this area and for permission to include it within these pages.

The staff members of Graffiti Publications have all contributed in some way to the production of this book and in particular I must make mention of the graphic design talents of Michael deWolfe.

In many ways the hot rodders of the world have contributed greatly to this book just by participating in the hobby. Many of the photos used as examples in this book were taken at street rodding events in Australia, New Zealand and the U.S.A. Even though the owners and builders of the cars included may not have known there handiwork would be so helpful to other rodders, I would like to thank them for their anonymous input on behalf of all hot rodders.

Finally I must thank my wife, Mary and my family for enduring the many late nights and times absent from home while this book was being written and published. I sincerely hope and trust that it will be a helpful guide to all hot rodders as they pursue their own dreams of building a safe and reliable street rod.

In this hobby there is no greater reward than to drive the street rod that you have built yourself.

Larry O'Toole

ADDITIONAL RECOMMENDED READING:

How to Make Your Car Handle, by Fred Puhn. Published by HP Books.
Chassis Engineering, by Herb Adams. Published by HP Books.
Boyd Coddington's How to Build Hot Rod Chassis, by Timothy Remus. Published by Motorbooks International.
Tex Smith's How to Build Real Hot Rods, by Tex Smith. Published by Tex Smith Publishing.
How to Build A Repro Rod, by John Thawley. Published by Steve Smith Autosports.
Carroll Smith's Nuts, Bolts, Fasteners and Plumbing Handbook, by Carroll Smith. Published by Motorbooks International.

MEASUREMENT CONVERSION TABLE:

Inches		Centimetres (10mm = 1cm)
.394	1	2.540
.787	2	5.080
1.181	3	7.620
1.575	4	10.160
1.969	5	12.700
2.362	6	15.240
2.756	7	17.780
3.150	8	20.320
3.543	9	22.860

Other titles from GRAFFITI Publications

Styling Street Rods
by Larry O'Toole.
Here's a complete guide to styling your street rod using other rodders project vehicles as a guide. Ten chapters and over 300 photos give an insight as to how other rodders have designed all aspects of their street rod including front and rear end treatment, engine bays, access and vision, interiors, running boards and fenders and even accessories.
$24.50

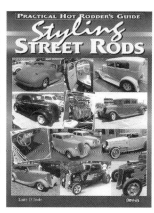

Nostalgia Street Rods
Our latest full colour production featuring the best in nostalgia street rods from all around the world. 112 pages in landscape format so you get to see the cars at their best with concise, accurate information and no through the spine photos.
$24.50

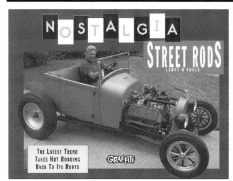

Street Flathead
By Mike Davidson.
This is the ideal book for the nostalgia hot rodder who seeks to power his or her street rod with a sweet running sidevalve. Everything you will ever want to know about building a high performance street flathead is contained in this book.
A follow-up his best selling Flathead Fever book with another great title for flathead Ford fans.
$18.95

A landscape format publication to show rods at their best in full living colour. No "through the spine photos" or close ups of hubcaps, just large format, glorious colour photos of your favourite street rods with short captions giving you all the basic information about each car.
$20.00

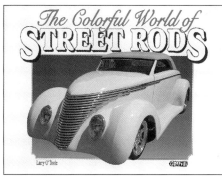

Flathead Fever
By Mike Davidson.
Detail the methods and tricks Mike has used to build two versions of the Ford Flathead engine. One is mildly modified for increased performance in a street driven hot rod, with occasional outings to the drag strip, while the second is an all out race engine for the salt flats, where Mike's knowledge and ability with this engine has been proven with speeds in excess of 160 m.p.h.

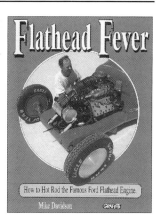

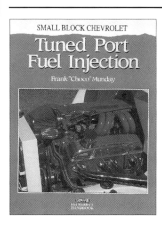

Small Block Chevrolet Tuned Port Fuel Injection
By Frank 'Choco' Munday
'Choco' Munday's experience really shows through with simple and logical, yet comprehensive detail. An extensive content list makes it very easy for the reader to quickly find the section dealing with any particular aspect of the TPI system. Combine this with extensive lists of diagrams, tables and photos and you have a very complete package that will impress even the most knowledgable Tuned Port Injection mechanic.

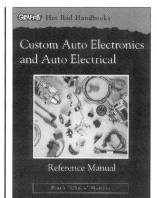

Custom Auto Electronics
By Frank 'Choco' Munday
Here's a book that tackles the electronic world in language that we can all understand and from a viewpoint of the hands-on enthusiast who wants to work with it on his own hot rod project. The book also covers conventional wiring including the fitting of an entire loom into a hot rod type vehicle.
Custom Auto Electronics and Auto Electrical Reference Manual is extensively cross-referenced to make your electrical and electronics research easy.

ALL BOOKS AVAILABLE FROM:
GRAFFITI PUBLICATIONS (Australia). TELEPHONE: 61 3 5472 3653 FACSIMILE: 61 3 5472 3805
CELEBRITY BOOKS (New Zealand). TELEPHONE: 64 9 486 0620
MOTORBOOKS INTERNATIONAL. TELEPHONE: 1 800 826 6600